CW00547870

CONVAIR
ADVANCED DESIGNS

Robert E. Bradley

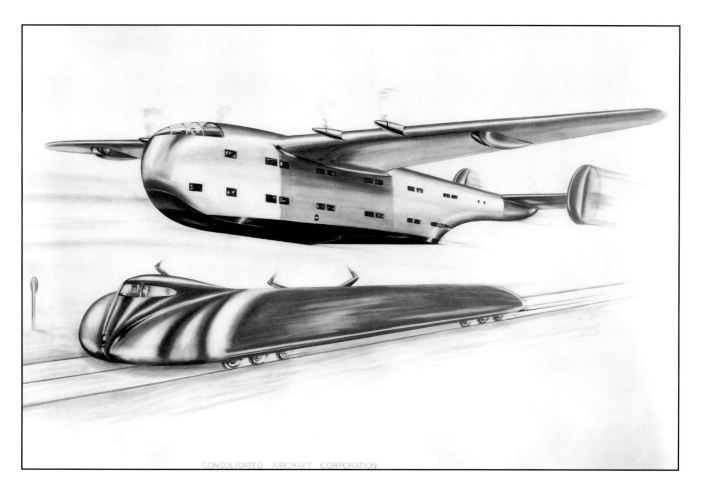

CONSOLIDATED AIRCRAFT CORPORATION

specialtypress
PUBLISHERS AND WHOLESALERS

Specialty Press
39966 Grand Avenue
North Branch, MN 55056
Phone: 651-277-1400 or 800-895-4585
Fax: 651-277-1203
www.specialtypress.com

© 2010 by Robert E. Bradley

All rights reserved. No part of this publication may be reproduced or utilized in any form or by any means, electronic or mechanical, including photocopying, recording, or by any information storage and retrieval system, without prior permission from the Author. All text, photographs, and artwork are the property of the Author unless otherwise noted or credited.

The information in this work is true and complete to the best of our knowledge. However, all information is presented without any guarantee on the part of the Author or Publisher, who also disclaim any liability incurred in connection with the use of the information.

All trademarks, trade names, model names and numbers, and other product designations referred to herein are the property of their respective owners and are used solely for identification purposes. This work is a publication of Specialty Press, and has not been licensed, approved, sponsored, or endorsed by any other person or entity.

Edit by Mike Machat
Design by Monica Seiberlich

ISBN 978-1-58007-133-8
Item No. SP133

Library of Congress Cataloging-in-Publication Data

Bradley, Robert E.
  Convair advanced designs : secret projects from San Diego, 1923-1962 / by Robert E. Bradley.
      p. cm.
  Includes bibliographical references and index.
  ISBN 978-1-58007-133-8
  1. Convair airplanes—History—20th century. 2. Research aircraft—History—20th century. I. Title.
   TL686.C58.B73 2010
   623.74'6—dc22
                                            2009042182

Printed in China
10 9 8 7 6 5 4 3 2 1

Front Cover:
The past meets the future, circa late-1950s, as a Convair Model 23B nuclear-powered seaplane bomber and mine layer is seen over an inboard profile drawing of a 1930s-era Consolidated long range "flying battleship" complete with an Admiral standing atop a commander's bridge ahead of the aircraft's wing. Plan view drawing of canard configuration Mach 2 bomber is seen at bottom. (Cover design concept and jet seaplane model courtesy of John Aldaz)

Title Page:
Only the artwork survives for this novel late-1930s concept for launching a 100-passenger flying boat at a much higher gross weight than would have been possible for a water takeoff. The streamlined "art deco" high-speed rail device was to function much like research rocket sleds used to only a few decades later, propelling the aircraft to high speed in a relatively short distance. (Convair via SDASM)

Front Flap:
Designed for Anti-Submarine Warfare missions, this tri-motor flying boat was designed to utilize dunking sonar as a primary detection method. Designated as the P6Y, the concept aircraft's high wing was mounted on two large pylons above the hulled fuselage reminiscent of another famous Consolidated aircraft, the PBY Catalina of World War II fame. (Convair via SDASM)

Back Cover Photos

Top:
Convair's mammoth six-engine XC-99 experimental cargo transport was the world's largest airplane when it first flew in 1947. It is shown here with the L-13, the company's entry into a USAF light observation aircraft competition. Less than a decade later, Convair production lines in Southern California and Fort Worth, Texas, were producing bombers, fighters, airliners, and even guided missiles. (National Archives via Dennis R. Jenkins)

Bottom Left:
Convair engineers faithfully utilized large scale models to test such operational attributes as hull architecture, configuration studies, and even loading of troops and cargo on a mock beachhead—even if that was only a bay shoreline in San Diego! Note the scale-model tank driving down the bow loading ramp. (Convair via SDASM)

Bottom Right:
Life imitates the art of modeling as a real Convair R3Y-1 Tradewind turboprop flying boat disgorges troops and an Army Jeep on Convair's mock Naval Base loading dock in San Diego. The large nose section raised a full 90 degrees to allow unobstructed access to the bow loading door. (National Archives via Dennis R. Jenkins)

Distributed in the UK and Europe by
Crécy Publishing Ltd
1a Ringway Trading Estate
Shadowmoss Road
Manchester M22 5LH England
Tel: 44 161 499 0024
Fax : 44 161 499 0298
www.crecy.co.uk
enquiries@crecy.co.uk

CONTENTS

Acknowledgments 4
About the Author 4
Introduction 5

Chapter One: About the Company 7

Chapter Two: Seaplane Programs 13
1. TW-3 14
2. NY-1 Husky 14
3. NY-2 Husky 15
4. O-17 Courier 16
5. Fleet Trainer 16
6. XPY-1 Admiral 17
7. Commodore 18
8. Fleetster 19
9. P2Y-1 and P2Y-2 Ranger 20
10. XP3Y/PBY Catalina 21
11. Model 28 Military Studies 23
12. Model 28 Commercial Studies 25
13. PB2Y Coronado 27
14. Model 29 Military Studies 28
15. Model 29 Commercial Studies 29
16. XPB3Y-1 30
17. Model 34 Commercial Studies 33
18. Navy Flying Wings 34
19. Trans-Oceanic Flying Boat 36
20. 100-Passenger Seaplane 38
21. Prewar Two- and Three-Engine Flying Boats 40
22. Prewar Four-Engine Flying Boats 43
23. Model 31 46
24. Model 31/XP4Y-1 Corregidor and
    Military Studies 48
25. Model 31 Commercial Studies 51
26. PJY Utility Seaplane 53
27. Generalized Seaplane Study 55
28. Postwar Seaplane Designs 58
29. Generic Four-Engine Research Model ...62
30. Model 37 Seaplane Study 64
31. P5Y 65
32. Long Range Transport Flying Boat 69
33. Assault Seaplane Transport 70
34. Long Range Special Attack Airplane 71
35. Cudda 73
36. Betta 75
37. High Performance Flying Boat 78
38. Twin Hull Seaplane 81
39. R3Y Tradewind 82

40. Supersonic Attack Seaplane 86
41. Seaplane Logistic Transport 89
42. Nuclear Seaplane 91
43. Nuclear Tactical Applications 95
44. XP6Y-1 99
45. Combat Seaplane 102
46. Assault Transport Seaplane 105
47. ARDC Seaplane Study 107
48. Princess Nuclear Test Bed 109
49. ASW Studies 110
50. Mach 4 Attack Seaplane......................113
51. ASW GETOL 115
52. Flying Submersible 117

Chapter Three: Bomber Programs 120
53. Guardian 121
54. Army Flying Wings 122
55. Blended Wings 127
56. Prewar One-, Two-, and
    Three-Engine Bombers 128
57. Prewar Four-Engine Bombers 136
58. B-24 Liberator 138
59. C-87 Liberator Express and Model 32
    Transport Studies 139
60. Model 32 Commercial Studies 142
61. Pre B-32 Studies 144
62. B-32 Dominator 145
63. Model 33 Transport Studies 150
64. Model 33 Commercial Studies 152
65. Model 35 153
66. B-36 Peacemaker 154
67. XC-99 157
68. Flying Wing B-36 Comparison 161
69. PB4Y Privateer 163
70. Tailless Twin-Engine Patrol Studies ...165
71. Conventional Twin-Engine Patrol Study ...168
72. Tailless Four-Engine Army Bomber 169
73. Twin-Engine Attack Airplane 171
74. XB-46 Jet Bomber 173
75. Turboprop Heavy Bomber 175
76. High Performance Patrol Landplane ...177
77. Carrier-Based Bomber 178
78. Carrier-Based Landplane 180

Glossary 183
Selected Source Material 183

ACKNOWLEDGMENTS

If this was an all me, me, me book I would say so, but it's not and not by a long shot. I do, therefore, want to say thank you and express my appreciation to those who were instrumental in seeing that this book was published and published right. First I want to thank my Editor Mike Machat (and Nick Veronico before Mike) for his guidance in keeping me on the straight and narrow and cleaning up my shortcomings, and to Specialty Press for making it all possible. I especially want to thank Dennis Jenkins for his original suggestion for the book while he was on a visit to the San Diego Air & Space Museum (SDASM), and for his inspiration and support to bring this book about. I would also like to thank all of my previous co-workers at General Dynamics, the SDASM Archive and Library Staff, and fellow volunteers with whom I have consulted, and especially Mark Aldrich whose encyclopedic knowledge of Convair aircraft and aviation history has been invaluable.

This book would not have been possible without the support and assistance of the San Diego Air & Space Museum and particularly Katrina Pescador, Head Archivist extraordinaire, for making the source material available for preservation and distribution in the form of this book. I also want to thank Scott Lowther of Aerospace Projects Review for making certain drawings available, as noted. In addition, there are photographs from the National Archives (NARA) courtesy of Dennis Jenkins, and design and editorial input from John Aldaz, for which I also want to offer my thanks. All of the graphics, with the exception of those noted, are from the SDASM.

Last, but not least, I want to thank my wife Linda (my in-house copy editor) for her patience, occasional forceful encouragement, and especially for protecting me from my literary and grammatical boo-boos.

After spending 40 years in the aerospace industry, I had seriously wondered if I really wanted or needed to do all the work this book involved, but in the end it turns out to have been very, very worthwhile to me, and I hope also to you, the reader.

ABOUT THE AUTHOR

Author Robert E. Bradley enlisted in the U.S. Army Air Forces immediately after World War II, serving in the Communication Service. After leaving the Air Force he attended the University of Southern California and graduated in 1953 with a B.Sc. in Physics. After graduation Bradley worked for North American Aviation in El Segundo, California, and in 1957, moved to Convair Astronautics (later General Dynamics Space Systems Division) in San Diego, where he remained until his retirement in 1993. Working initially in Operations Analysis in Advanced Engineering, he moved to the Economic Analysis Group until that group was integrated into the Contracts and Estimating Department. After his retirement in 1993 as Manager of Economic Analysis, he became a volunteer archivist at the San Diego Air & Space Museum, specializing in the Museum's Convair archives and space and missile collection. Bradley has a continuing long-standing interest in aerospace history and the advanced design studies and proposals produced by Convair's Preliminary Design Departments. He lives with his wife Linda in San Diego, California, and both are avid photographers and seasoned world travelers.

I n the past, books of the aviation genre have had a rich history of programs and aircraft experience from which to draw. And authors have done just that. Indeed, thousands of books have been written about all manner of aircraft and the people who built and flew them, the technology, and the operational use of those airplanes. One of the results of the abundance of this subject material is that minimal attention was given to those aspects of advanced engineering such as design studies and proposals. In fact, with the exception of a few notable authors, little attention was paid this subject matter and in many cases dismissed as of no importance or undeserving of attention.

This view has started to change. One of the reasons is that the economic realities of technology's ever-increasing complexity and sophistication result in fewer and fewer programs being undertaken. The large variety of aircraft development and production of the past is rapidly decreasing, and with it, the subject material for the aviation history author. In recent years, however, there has been a significant interest developing for the technical niche of advanced designs and the "planes that didn't happen." Since this development of interest, more and more books are appearing that address this aspect of aviation engineering. It is my hope that this book makes a small contribution to the trend of preserving this previously neglected aspect of aerospace history.

This is a book detailing advanced design studies and proposals conducted by Consolidated Aircraft Corporation (CAC), and later Consolidated Vultee Aircraft Corporation (CVAC, informally called Convair) and General Dynamics Convair Division at its San Diego plant, GD-SD (collectively referred to as the Company). These are airplane designs that never made it past the preliminary design phase. Some were competitive proposal responses to customer solicitations for specific airplane requirements. Some were exploratory design efforts to permit the Company to not only stay abreast of technology but to be an innovative leader in the aircraft arena. Many of these studies were funded by the customer and some were Company sponsored. Others were marketing efforts intended to interest the customer in advanced versions or alternate uses of ongoing aircraft programs.

The time period of the studies and proposals, and completed aircraft projects, included in this book run from the early 1920s to the early 1960s, which is the period covered by the documentation available in the San Diego Air & Space Museum's archives. This book covers primarily those projects initiated at the San Diego location. Several of the major programs herein had their start in San Diego, but were then transferred to Consolidated's Fort Worth, Texas, facility. Follow-on activities associated with these programs accomplished at Fort Worth are not included, nor are any of the newly proposed projects that originated in Fort Worth.

This volume covers seaplanes (Chapter Two) and bomber designs (Chapter Three) originated at San Diego, and the previous Buffalo, New York, facility, presented in generally chronological order. Although the main emphasis here is the advanced designs produced by the Company, some of the related aircraft actually built during this period are briefly noted to provide chronological context to the discussion of the subject projects. Transport versions of the mainline combat aircraft are also discussed. This approach was taken because of the integral nature of both designs and the overall program context. In the case of seaplanes, however, the fighters, including the SeaDart and Skate, are not discussed because they were stand-alone fighter programs.

Although the Company's work on real-world proposals are hoped to bear fruit, many of the design studies were aimed at establishing possibilities and in advancing technical design trends.

The source of the majority of the material for this book is the archival material donated to the San Diego Air & Space Museum in 1995 by General Dynamics Corporation when the Convair Division in San Diego was liquidated. Unfortunately, the material is very limited and fragmentary, since the Company did not have a dedicated and organized system for the preservation of historical material. Of course, during the Cold War, security was a prime concern and, as is believed occurred with many organizations (the government was just as guilty), it was much easier to destroy documentation than to declassify it. This tends to severely limit the historical documentation available from that time period. The end result is that only a small fraction of the Company's technical history survived. Further, most of the material that did survive was only due to a small group of dedicated individuals who had a passionate interest in the Company's technical history and legacy. Three that come to mind are Gordon Jackson, Howard Welty, and Bill Chana.

The nature and physical quality of the archival material varies considerably, which is not surprising in that the earliest documents are from the mid- to late-1930s. Most of the documents and drawings are reproductions, and there is very little original material. In those early time periods, reproduction technology was quite primitive, certainly by today's standards. Most of the earliest brochures consisted of photographic prints of illustrations or drawings and what is believed to be ozalid reproductions of data, graphs, text, and often drawings such as three views, which appear especially poor and tended to fade. Some photo reproduction is of acceptable quality but tends to be at the mercy of photographic exposure. Also, reduced-size drawings often have lines too thin to reproduce well.

Interestingly, color images included in brochures were all watercolor-tinted photographs, and painted individually, indicative of the lack of color reproduction at that time. In many cases the drawings were done in pencil in the final size, for instance three views in 8½x11-inch report size. The earliest illustrations and artwork range from the quite primitive to fairly good. There seemed to be a lack of imagination in the 1930s and 1940s and the illustration depictions were fairly stereotyped in that the same view and angle was shown for each of the projects, lending an undeserved amount of sameness to the presentation.

The shortcomings of these drawings and illustrations may be seen in some of the images here, especially in the case of second- or third-generation reproductions. Some lesser-quality images are included, however, because of their importance in documenting the subject. It wasn't until the postwar period, when the reproduction process using photographic negatives became prevalent that much higher quality drawing images began to be seen. By the 1970s, photocopy and digital reproduction revolutionized these processes. All of the drawings included here are originals from the archives, and none have been redrawn.

As noted above, the surviving documents held in the Museum's archives represent only a fraction of the Company's technical history. While there are some excellent brochures from the 1930s and 1940s, there are also letters, drawings, and photographs that represent fragmentary glimpses of projects and activities, but no overall coherent description or definition has survived. One can only surmise the circumstances and motivations for most of the projects involved. When dates are quoted, they are generally from report or brochure publications indicating a time of completion, or from dates on drawings, photographs, and illustrations.

Chapter One

About the Company

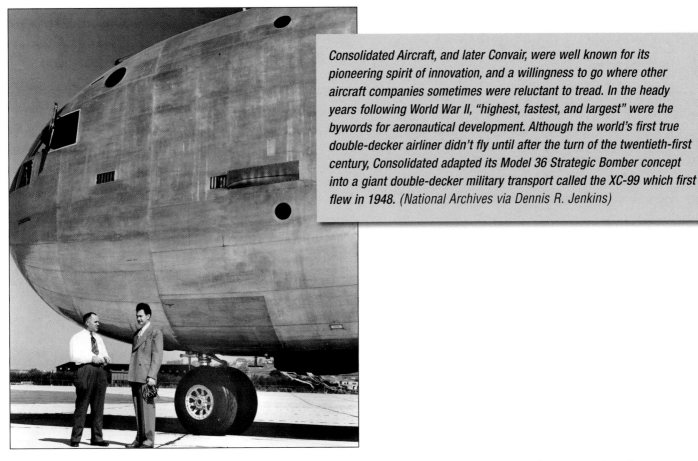

Consolidated Aircraft, and later Convair, were well known for its pioneering spirit of innovation, and a willingness to go where other aircraft companies sometimes were reluctant to tread. In the heady years following World War II, "highest, fastest, and largest" were the bywords for aeronautical development. Although the world's first true double-decker airliner didn't fly until after the turn of the twentieth-first century, Consolidated adapted its Model 36 Strategic Bomber concept into a giant double-decker military transport called the XC-99 which first flew in 1948. (National Archives via Dennis R. Jenkins)

About the Company

This short history encompasses the origins, background, and chronology of Convair's San Diego Division. It is hoped that this will provide the context and overall perspective to better appreciate the evolution of the technical design history presented herein.

In the birth and early evolution of the airplane, individuals built the industry as inventors in barns and makeshift workshops that were to become the industrial giants. In its earliest days, aviation was driven by an inspired, inventive, and dedicated breed of men. The pioneering accomplishments of Curtiss and Martin and later Douglas, Boeing, and many others, come to mind. In many of these examples the intensity, dedication, passion, and leadership of an individual brought it all about. In the case of Consolidated Aircraft Corporation that man was Reuben H. Fleet.

Reuben Fleet was a successful real estate businessman living in Montesano, Washington, in 1917. He had served as a member of his state legislature, was a Captain in the National Guard, and was a dedicated aviation enthusiast. As World War I became inevitable in his eyes, he was able to obtain an appointment to the aviation branch of the Army Signal Corps at the age of 30. He left home in March 1917 to report for duty at the Aviation Section flight school at North Island in San Diego, California. Fleet had some acquaintance with San Diego when he served a short period patrolling the Mexican border with the National Guard in 1911. On 5 April 1917 he took his first flight, just one day before the declaration of war. On 9 August 1917 he graduated and received his wings.

After graduation, now Major Fleet spent several months as Acting Commanding Officer Eighteenth Aero Squadron-Training, in charge of the North Island machine shop. Early in 1918, Fleet was assigned to the Signal Corps aviation headquarters in Washington,

D.C., as Executive Assistant to the Chief of Training. In this assignment he was principally involved in establishing pilot training schools.

On 3 May 1918, Major Fleet was assigned by the Air Service to organize and inaugurate a U.S. Air Mail Service in conjunction with the Post Office Department. As Officer In Charge, the service was initiated on 15 May 1918 and the Army operated it until 21 August when the Post Office Department assumed the operations and control.

Major Fleet returned to the Air Service Training Department and in June was assigned to open the Mather Field advanced training facility in Sacramento, California, where he spent three months until September 1918. Upon returning to Washington, D.C., he arranged a trip to England in mid October to observe British pilot training methods. While there the armistice took place and the war was over on 11 November 1918.

Fleet returned from Great Britain in late December and requested to remain in the Army and be assigned to the Engineering Division at McCook Field, Ohio, hoping to be involved in training airplane development. When he arrived in January 1919, he found McCook Field to be particularly chaotic and disorganized, as could be expected during the transition to peacetime status. Major Fleet soon became the Business Manager and Contracting Officer for McCook Field. He remained there until he resigned in November 1922, after more than four years at the Air Service Engineering Division, to return to the civilian sector. Fleet then took a position with Gallaudet Aircraft Corporation in East Greenwich, Rhode Island, as Vice President and General Manager, but moved on soon thereafter to organize a new company of his own.

When Major Fleet formed the new company, Consolidated Aircraft Corporation (CAC), on 23 May 1923, he did so from the remnants of the Gallaudet Aircraft

Corp. and the Dayton-Wright Company, both of which were in serious financial difficulties. Fleet, working for Gallaudet at that time, recognized it had slim prospects for the future. The Dayton-Wright Company in Dayton, Ohio, had been a large airplane producer during World War I and had been purchased by General Motors in 1919, but also had few prospects for new business.

Major Fleet's special interest in training airplanes and pilot training had developed during his service with the Air Service Training Department. One of Dayton-Wright's projects at that time was the TW-3 training airplane for the Army and it had built a prototype. The Air Service was interested in buying a quantity of the TW-3 trainers and Fleet moved on this opportunity. In May 1923, he negotiated with Gallaudet to lease its factory facilities and to use its work force. A short time later Fleet bought the TW-3 trainer, its design, and certain patents from Dayton-Wright. A production order for 20 TW-3s was forthcoming and was to be the first aircraft for the new Consolidated company, and the birth of an aviation giant.

After successfully completing the TW-3 production contract, Consolidated won a competition for a follow-on trainer, the Army's PT-1. Because of a restricted factory facility and limited workforce availability at Gallaudet in East Greenwich, Fleet started looking for a new location for the company. In a short time he settled on Buffalo, New York, because it had a good skilled labor supply and available factory space. The facility itself, of which Fleet leased a portion, was the Curtiss factory used for the production of aircraft in World War I. The lease was finalized on 22 September 1924 and the move was made shortly thereafter.

Consolidated's trainer business was very successful, both with the Army and the Navy. A total of 847 trainers were produced through 1930, including the Army's PT-1, PT-2, and PT-3, and the Navy's NY-1, NY-2, and NY-3.

When Major Fleet formed his company in May 1923, his first contract was to build 20 TW-3 Army trainers. He immediately leased space at the troubled Gallaudet Aircraft factory in East Greenwich, Rhode Island, where he had been working, and employed the Gallaudet workers to successfully complete this first Consolidated Aircraft project. (Convair via SDASM)

Looking to expand its market to larger airplanes, and after a failed joint venture with Sikorsky in 1927 for an Army Night Heavy Bomber competition, Consolidated became interested in a Navy requirement for a new long-range patrol bomber flying boat. The Company initially won the competition for the prototype XPY-1 in February 1928 but lost the subsequent production contract. Undaunted, Consolidated continued to work on an advanced version of this aircraft and was indeed successful in gaining a production contract for a follow-on P2Y-1 and P2Y-2 Ranger seaplane program in 1932.

Early in 1928, Major Fleet embarked on a civilian trainer project, a scaled down version of the military trainers being built and initially named the Husky Jr. The first of these trainers, subsequently renamed the Fleet, was built by Consolidated, but the Corporate Board of Directors was concerned that the Company could not build a commercial airplane economically in a military manufacturing environment. This led to the sale of the trainer to Major Fleet, who set up a new company, Fleet Aircraft of Canada, Ltd., nearby in Canada to produce the trainer. Consolidated's Board then began to see the potential for this airplane and six months later repurchased it from Fleet. Consolidated went on to produce 437 Fleets and 192 equivalent airplanes in spares and parts. The Canada facility continued to build the Fleet until it was sold to Canadian interests in 1937.

Thomas Brothers Aircraft Company was formed in 1912 and located in Ithaca, New York, in 1914. Two English brothers formed it, W. T. Thomas (who had worked for Curtiss) and Oliver Thomas. By 1916, they were under-financed and at risk when Frank Morse and Herman Westinghouse stepped in and formed the Thomas Morse Airplane Company in January 1917 as a unit of the Morse Chain Company. The airplane

company was quite successful and they built approximately 600 airplanes during the war years, including the famous Tommy Scout trainer. The postwar years were difficult but they did win an Army competition for an all-metal version of the Douglas O-2 observation aircraft and successfully built six of the aircraft. The Army was, however, skeptical about Thomas Morse's ability to carry out a production program for this airplane. Major Fleet was contacted by the Army to assess his interest and the end result was a successful acquisition of the company in August 1929. Thomas Morse was moved to Buffalo and integrated as an operating entity of Consolidated.

The Navy initiated a competition to replace the P2Y Rangers in 1933. The finalists were Consolidated and Douglas, which were both awarded prototype contracts. In the end, Consolidated won with its XP3Y-1, to be redesignated the PBY, which was built in Buffalo and had its first flight in March 1935. The first order for 60 production aircraft came quickly in June, and Consolidated's most successful seaplane program was underway.

Two of the problems that Consolidated encountered with the Buffalo location were the poor winter flying weather and the lack of open water for flight of the seaplanes. Fleet had been surveying alternate factory locations for several years and had looked at potential sites on both coasts. While on a countrywide Fleet trainer sales trip in August 1929, Major Fleet actually made an offer to the City of San Diego to buy Lindbergh Field for $1 million, but nothing came of it.

After several years, the competition for the relocation site was narrowed to San Diego, Los Angeles, Long Beach, and a strong bid by Buffalo to retain the company there. San Diego ended up the winner and the Board of Directors authorized a conditional lease on 29 May 1933. In December 1934 bids for the construction

Consolidated went on to win an Army follow-on procurement in mid-1924 for a new primary trainer, the PT-1. Fleet had been concerned about the availability of skilled workers in the Rhode Island location and was actively looking for a new home for his Company. He finally settled on Buffalo, New York, as a location with a good labor supply and available space in an existing Curtiss factory. Consolidated moved there in September 1924 and remained for 11 years. (Convair via SDASM)

of the first factory unit were requested. Construction started soon thereafter on the $300,000 factory building. The move from Buffalo required only a short period of production down time, from mid-August 1935 to shortly after Labor Day in September. The dedication of the new 275,000-square-foot facility took place on 20 October 1935.

Consolidated had been awarded two large contracts, the first for 50 P-30 (PB-2A) fighters taken over from the financially troubled Detroit Aircraft Company in 1931. The second contract was an order for 60 PBY seaplanes received just three months before the new plant's opening. The company found itself short of manufacturing space almost as soon as the first factory unit was completed. Another building was initiated almost immediately in a further plant expansion, and by 1937 the floor space totaled 450,000 square feet.

As the war loomed and mobilization was underway, burgeoning orders for PBY seaplanes and the newly initiated B-24 program created even more demand for factory space. The main plant at Lindbergh Field, Plant 1, had been nearly doubled in size and in the spring of 1941 was doubled again with two large buildings (Buildings 2 and 3) totaling an additional 646,000 square feet. In November 1940, construction was started on a huge new factory, Plant 2, with 1.593 million square feet. This new production facility was about 1/2

mile north of Plant 1. Plant 2 was dedicated on 20 October 1941, doubling again the total factory space of Consolidated in San Diego.

Major Fleet relinquished his position as Chairman and President of Consolidated Aircraft Corporation at the end of 1941. He had a strong aversion to President Roosevelt's new tax laws that were to take effect in 1942 and felt the time was right to sell his 34-percent interest in the company. The buyer was the Aviation Corporation (AVCO), a holding company headed by Victor Emanuel, and the sale was finalized on 24 November 1941. Among the AVCO companies were Vultee and Stinson. The formal merger of Consolidated and Vultee took place 17 March 1943 and the new entity was named Consolidated Vultee Aircraft Corporation (CVAC). The company became known as Convair but that name was not formalized until 1954.

Aircraft production at the San Diego Division during the war years reached a total of 2,160 PBYs, 6,724 B-24s, and 210 PB2Ys. The wartime employment of the Division peaked at 39,500 people in 1943. "Nothing Short Of Right Is Right" was Reuben Fleet's popular slogan used by Consolidated to instill pride of workmanship in its thousands of employees.

After the merger with Vultee, the Consolidated Vultee Aircraft Corporation had 13 divisions in 10 states and a total of 101,600 employees.

Consolidated had conducted a very successful trainer business and was also able to enter the seaplane market with production orders for the Commodore and the P2Y Ranger program. A serious disadvantage for seaplanes at the Buffalo location was the winter freezing of the Niagara River. An extended survey was conducted to again move the Company, and San Diego, California, won out. After building a new 275,000-square-foot factory by San Diego Bay, the move was carried out in August 1935 and the dedication was held on 20 October that year. (Convair via SDASM)

During World War II, Convair employment peaked at 40,000 workers who built 9,094 airplanes during the war years. More than 40 percent of the employees were women, and the legendary term "Rosie the Riveter" originated here. Fleet's famous slogan "Nothing Short Of Right Is Right" was used to instill pride of workmanship and wartime quality in the workers. (Convair via SDASM)

After World War II ended, the aviation industry overall went through a severe but expected contraction. By 1946, Convair was down to five operating divisions: San Diego, Fort Worth, Vultee Field, Stinson, and Nashville. The San Diego Division reached a low point of employment of 3,400 in 1945.

There was active consideration in the second half of 1946 of merging Consolidated Vultee into the Lockheed Corporation. Significant discussions took place and in-depth analyses were undertaken by both companies as to benefits, the manner of implementation, and the effects such a combination would have on each of the organizations. The prospects for this merger then evaporated, reportedly because of the problems Lockheed was having with its Constellation airliner.

Floyd B. Odlum, President of the Atlas Corporation, an investment house, became interested in the potential of Consolidated Vultee and in 1947 acquired seven percent of AVCO stock. In November, Atlas Corporation became the controlling stockholder when AVCO agreed to withdraw in order to manage its Nashville facility and the ACF Brill Company. Vultee was then closed and its activities moved to San Diego, and some time later the Stinson assets were sold to Piper Aircraft. On 15 May 1953, controlling interest in Consolidated Vultee moved into the hands of John J. Hopkins, the Chairman of General Dynamics Corporation, after that company acquired sufficient stock to become the majority stockholder. General Dynamics was on its way to becoming an industry powerhouse. Convair, a Division of General Dynamics, was officially the new name as of 30 April 1954.

The San Diego Division's mainline postwar peacetime projects included the modestly successful 240-340-440 medium airliners and a variety of prototype aircraft programs such as the B-46, P5Y, R3Y, FY-1, F2Y, and F-92A. It wasn't until the advent of the Korean War and the Cold War with the Soviet Union, starting in the 1950s, that Convair started to make a comeback. The principal programs during this period were the F-102 and F-106 interceptors and the Atlas ICBM, and with employment peaking at 32,000 in 1956. The Atlas program was split off and a new Astronautics Division was formed in 1954, and employment in that division peaked at 31,000 in 1961.

In 1956, Convair San Diego undertook development of a jet transport, the 880/990 that unfortunately became a financial disaster. As it turned out, it was the wrong airplane at the wrong time and Convair was to only build 102 of these aircraft. The company ended up losing about $425 million on the venture.

The Division continued to decline with the end of F-106 production in 1961 and no further aircraft were produced. In 1965 the Astronautics Division was again merged with Convair San Diego. Convair's business was maintained as a structure subcontractor, building the empennages for the Lockheed C-141 and the C-5 transports. In 1969 Convair won an award from McDonnell Douglas to build fuselage sections for the DC-10 and later for the MD-11 and KC-10 that lasted until 1996. The San Diego Division and the Fort Worth Division were combined for a short period from 1971 to 1974 as the Convair Aerospace Division, but this proved unworkable. The one bright spot for Convair was the winning of the Tomahawk program in 1974, a program that proved to be quite difficult but eventually turned out to be very successful. The space activities were again split off from Convair with the establishment of the Space Systems Division.

The fall of the Berlin Wall in 1985 signaled the end of the Cold War with the Soviet Union and introduced the specter of sharply reduced defense budgets available for the aerospace industry in the future. General Dynamics, under the leadership of CEO (and former Apollo astronaut) William A. Anders, recognized the inevitable consolidation that was in store for the industry and was very proactive in deliberately carrying out future planning. Early on, GD evaluated both the acquisition route and the divestiture route as approaches to the future of the Corporation. Because the entire process of industry consolidation was just getting underway, and although significant defense budgets were still available at the time, it still appeared there was limited opportunity to grow through the purchase of companies or divisions. GD then decided on the divestiture option and subsequently sold all of its aircraft, space, and missile divisions.

An agreement was reached with the Martin Marietta Corporation in 1993 for the sale of the Space Systems Division, the home of the Atlas for 45 years. The sale of that division closed 1 May 1994. The Tomahawk cruise missile program was also sold to the Hughes Aircraft Company in 1994. The San Diego Convair Division reached an agreement with McDonnell Douglas in 1994 to terminate the contract for fabrication of MD-11 fuselage sections and to transfer that work back, and Convair's subcontracting activities ceased. The last fuselage structure, No. 166, was delivered in early 1996.

The entire General Dynamics Convair San Diego Division was closed in 1996 and the Lindbergh Field Convair facility, Plant 1, was demolished shortly thereafter. Plant 2 was taken over by the Navy's Space & Naval Warfare Systems Command (SPAWAR). One of the three Plant 2 buildings, however, remained leased to Lockheed Martin for continued fabrication of the Atlas and Centaur space vehicles' propellant tanks, an activity that had been located there originally.

Reuben Fleet brought Consolidated Aircraft to San Diego in 1935 from Buffalo, New York, and for half a century, Convair was the largest civilian employer in San Diego. Employment peaked in 1943 at 40,000 employees, consisting of 40-percent women. The term "Rosie the Riveter" was originally coined at Consolidated. Consolidated and Convair were responsible for building some of the most significant aircraft and missiles in the history of aerospace, including the PBY Catalina, B-24 Liberator, F-102 and F-106, the first delta-winged interceptor, the Tomahawk cruise missile, as well as the Atlas, and the Atlas Centaur ICBM and space vehicles.

Prior to the move in 1935, Consolidated had just received two large contracts for P-30 fighters and for its new PBY-1 follow-on Patrol Bomber. Almost immediately after the move, the company became cramped for space and launched an expansion that doubled the factory floor space. With the war preparedness efforts in the early 1940s this location, Plant 1, was again doubled in size with an addition (center) and a new Plant 2 was built about 1/2 mile north, again equal in size to the initial Lindbergh Field location. This aerial photo was taken in the early 1960s at the peak of production for Convair's 880 and 990 jetliners visible on the ramps. With the demise of Convair in 1996, Plant 1 was demolished, but Plant 2 is still in use. (Convair via SDASM)

CHAPTER TWO

SEAPLANE PROGRAMS

The final result of numerous predesign studies and a prototype test program (P5Y) evolved into the largest turboprop flying boat ever used by the U.S. Navy. Convair's R3Y-1 Tradewind set new standards for sea-going heavy-lift transportation. The Tradewind was capable of carrying men and munitions across the oceans and deploying them into battle on beachheads or ports anywhere in the world within 24 hours. The actual airplane you see here would not have been possible without endless concept study and development. (National Archives via Dennis R. Jenkins)

Seaplanes, airplanes that have the ability to takeoff and land on water, have been of interest to designers since the earliest days of aviation. The prospect of over-water flying and operating from waterways where no airports exist was a big attraction of the idea of seaplanes. In addition, the allure of less-restricted takeoff and landing direction and relatively fewer obstacles for these takeoffs and landings balanced some of the disadvantages and risks of water operations.

Glenn Curtiss conducted the first takeoff and landing from water in the United States in San Diego, California, on 26 January 1911. He accomplished this with his hydro-aeroplane, a Curtiss Model D, modified with the addition of floats. Glenn Martin followed with a similarly configured hydroplane in Newport Harbor, California, about one year later, in the spring of 1912. These pioneering flights were followed by many others in the early history of aviation, and the seaplane, as a significant aspect of the technology, was underway.

1 TW-3 (1923)

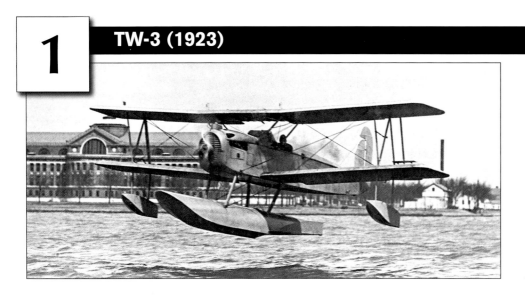

A seaplane version of the TW-3 appeared quite early in the trainer program. One of the TW-3s built by Consolidated for the Army was transferred to the Navy and was modified with the addition of floats and what appears to be a somewhat changed engine cowling and exhaust system. This aircraft became the first Consolidated seaplane. *(Convair via SDASM)*

When Major Reuben H. Fleet formed his new company, Consolidated Aircraft Corporation (CAC) in 1923, he took over the Dayton-Wright Company and the Gallaudet Aircraft Company, both in serious financial trouble. CAC's seaplane experience in the earliest days was limited to float-equipped versions of their training and small airplane stable of vehicles. Their first aircraft was the TW-3 that had been designed at the Dayton-Wright company. The aircraft was a two-place biplane designed by Virginius Clark of Clark Y airfoil fame, who later went to work for Major Fleet. The TW-3 had a wingspan of 34 feet 9 inches, a maximum takeoff weight of 2,447 pounds, and was powered with the liquid-cooled 180-hp Wright E engine based on the wartime Hispano-Suiza design. Because of Fleet's general interest in training airplanes and being aware the TW-3 was of current interest to Army Air Service, he pursued this opportunity and received his first contract for the new company almost immediately, in June 1923. He used the former Gallaudet factory in Greenwich, Rhode Island, to build 20 TW-3 trainers for the Army. One of those aircraft was subsequently transferred to the Navy (S/N A 6730) and was modified and tested with a single main float plus wingtip floats and thus became the first Consolidated-produced airplane to operate as a seaplane.

2 NY-1 Husky - Model 2 (1925)

This NY-1 Husky is believed to be at Naval Air Station Anacostia, Maryland, possibly during its testing program. The NY-1 was essentially a PT-1 incorporating the Navy's Wright J-4 Whirlwind radial air-cooled engine and a somewhat larger empennage. The Navy procured both landplane and seaplane versions of the NY-1 but it is not known how many were actually float-equipped. *(Convair via SDASM)*

After the TW-3 program, Major Fleet was well prepared when the Army held a competition in the summer of 1924 for additional trainers that included five other manufacturers. The fledgling Consolidated prevailed in this competition and the Army quickly proceeded with the procurement of 50 airplanes in July. This airplane, designated the PT-1 and named the Trusty (CAC Model 1), had a wingspan of 34 feet 5½ inches, gross weight of 2,577 pounds, and was also powered by the Wright E engine. This design of the PT-1, heavily based on the TW-3, became highly successful and incorporated much of Fleet's interest and concern for a safe and forgiving training airplane. The airplane had an outstanding safety record especially when compared with the JN-4 and the JN-6 that it replaced.

The next year, in 1925, Consolidated entered a competition for the Navy's new training airplane and proposed a version of the Army's PT-1. Again Consoli-

dated won, this time against 14 competitors. The entry was designated the NY-1 Husky (CAC Model 2) and an initial order for 16 immediately followed. The NY-1 differed from the PT-1 primarily in that it incorporated the new Navy-sponsored 200-hp J-4 Wright Whirlwind radial air-cooled engine, and a slightly enlarged vertical tail surface. The first flight of the NY-1 took place on 12 November 1925. The aircraft was procured as a landplane and had provisions for a float arrangement to permit water operations similar to that tested on the TW-3. Many of the NY-1s were thus equipped.

3 NY-2 Husky - Model 2 (1926)

The NY-2 was contemporary with the Army's new PT-3 but it differed very little in appearance from the NY-1. It had a slightly increased vertical tail surface area and later the wingspan was increased to 40 feet, a modification that was retrofitted to many of the existing NY-2s. The NY-2 also used the Navy's new 220-hp Wright J-5 Whirlwind engine. (Convair via SDASM)

The trend to the radial air-cooled engine continued as the Army trainer evolved to the PT-3 version of the PT-1 with a 220-hp Wright J-5 (R-790-8) radial engine replacing the older water-cooled Wright E (Hispano-Suiza) engine. The Navy followed suit with orders for a follow-on NY-2 with the same Wright J-5

and later with a slightly larger wingspan of 40 feet. The enlarged wing was retrofitted on a number of earlier NY-2s. The NY-2, like the NY-1, was procured as both landplane and seaplane. Several NY-2 versions were produced, including NY-2A gunnery trainers and NY-3s with a more powerful 240-hp Wright engine.

4   O-17 Courier - Model 8 (1927)

Another seaplane version of this PT/NY series was a float-equipped O-17 Courier sold to Canada. The O-17 (CAC Model 7) was an advanced trainer modification of the PT-3 used largely by the National Guard. Three of this type were sold to Canada, one of which was float equipped (CAC Model 8).

Archival sales records show 827 of this overall series of military trainers were built which included 306 Navy NY-1s and NY-2/3s, 470 Army PT-1s and PT-3s, 29 Army O-17s, 21 foreign airplanes, and 1 commercial airplane.

In 1928 the Army ordered a small quantity of an O-17 (CAC Model 7) version of its PT-3 Trusty for training National Guard pilots. Consolidated also sold three of these O-17 to Canada. One of these airplanes was equipped with floats and received a new model number, CAC Model 8. (Convair via SDASM)

5   Fleet Trainer - Model 14 (1928)

The Fleet was a small trainer and sport airplane built for the civilian market based on a scaled down version of the Husky Navy trainer design. The first flight of this airplane, initially named the Husky Jr., took place on 9 November 1928. The Fleet was fitted with twin floats and was certified as a floatplane in October 1929, about one year after receiving its initial ATC. There is no record of how many of the approximately 500 Fleets were so equipped. (Convair via SDASM)

After Consolidated's quite successful venture into training airplane market Major Fleet started looking at the potential for civilian trainers. Until this time most of the private flying utilized surplus airplanes from World War I that were becoming much more difficult and costly to maintain because of the lack of spare parts. The Husky was considered as a candidate; however, it was a relatively large airplane and was, in all probability, too large and expensive for the civilian market. Consolidated's designers then set about to scale down the Husky to about 75 percent of its original size. The design was finalized for a Husky Jr. (CAC Model 14) that closely met the target empty weight of 1,000 pounds. Construction then proceeded, and three airplanes were built in 26 days. The first flight

took place on 9 November 1928 and the Approved Type Certificate (ATC) was received the same month. The airplane, the Model 14, was soon renamed the Fleet. An amended ATC was issued for a twin-float seaplane version on 29 October 1929. The Fleet initially sold for $4,985 but during the Depression the price was reduced by $1,000.

The first Fleets were manufactured in Consolidated's Buffalo plant but doubts on the part of the Board of Directors fostered a sale of the airplane and its rights to Major Fleet, who set up a factory under his own name close by in Canada. The Board then reconsidered, and repurchased the design about six months later. Company sales records indicate a total 437 Fleets were sold plus spares equivalent to 192 airplanes.

Seaplane Programs

Consolidated bid on and won a contract for an airplane to fulfill the Navy's new requirement for a groundbreaking long-range twin-engine patrol seaplane. The XPY-1 Admiral was powered with two Pratt & Whitney Wasp R-1340 engines, and had a wingspan of 100 feet. Consolidated ended up losing the production contract to Martin because of higher cost. This long-range patrol monoplane design essentially ended the large biplane era in the Navy. (Convair via SDASM)

Now well established as a successful supplier of trainers, Consolidated was intent on expanding its scope, and started looking at larger and multi-engine airplane opportunities. Lacking engineering experience with large aircraft, Consolidated teamed with Sikorsky in 1927 to compete for a twin-engine heavy night bomber for the Army. Unfortunately this project turned out to be an also-ran in the Army's competition because of design deficiencies and poor performance.

About the same time as the Army's heavy bomber competition, another promising opportunity was developing. The Navy was in the process of formulating specifications and requirements for a new, large, long-range twin-engine patrol aircraft. The proposal called for a monoplane using the latest air-cooled radial engines, Pratt & Whitney 425-hp Wasps, and was to have a 2,000-mile range and cruise at 110 mph.

Reuben Fleet had just hired I. M. "Mac" Laddon from the Air Service Engineering Division, and had moved the Dayton offices back to Buffalo, where they started the initial foray into the flying boat market. Laddon had multi-engine experience with Air Service at McCook Field and had also worked on flying-boat designs. The proposal paid off and they were awarded a contract for $150,000 on 8 February 1928, for the design and production of one prototype airplane, the XPY-1 (CAC Model 9). Construction of this Consolidated flying boat, named the Admiral, started in March. By December the airplane was ready to fly but the winter weather was a problem for flying boats in Buffalo, as the Niagara River generally froze. These conditions would plague Consolidated until the move to San Diego in 1936. The decision was made to ship the XPY-1 to the Naval Air Station at Anacostia, Maryland, where it made its first flight on 11 January 1929. After a series of very successful tests it was officially demonstrated for the Navy on 22 January 1929. The innovative XPY-1 design ended the era of the large seaplane biplane forever.

At that time, all of the military services were compelled to open the competition for the production phase of a particular airplane. This meant that the designer of that airplane didn't necessarily receive the production contract. This happened on the PY-1 procurement, initiated in June 1929, when Martin won the quantity production order for nine flying boats, much to the disappointment of Reuben Fleet. Fleet had invested about a half million dollars in engineering and pre-purchased material over and above the original contract value in anticipation of the production contract. When Consolidated attempted to recover that investment in the proposal, Martin was able to underbid. Not all was well for Martin, however. It turned out it had numerous problems building Consolidated's design that resulted in significant schedule delay. This in turn caused considerable customer dissatisfaction. Shortly after, the competitive bid system for production process was ended.

After Consolidated lost the production contract for this Navy patrol airplane the company found a promising market for a commercial transport version of the XPY-1. It then went on to build and sell 14 of these Commodore 22-passenger flying boats to the New York, Rio, and Buenos Aires Line for use on their Latin American routes. NYRBA was later merged with Pan American Airways (PAA) which used the Commodore for many years. (Convair via SDASM)

The Navy specifications for the XPY-1 included a requirement for convertibility to a 32-passenger transport. Reuben Fleet, mindful of this capability, decided to investigate commercial market possibilities. During this period, potential Central American and South American air transportation venues were starting to emerge for both mail and passenger routes. Early in this period Fleet became involved in helping to organize an airline company, the Tri-Motor Safety Airways, with James K. Rand, Jr., of the Remington Rand Company and Capt. Ralph O'Neill.

Fleet's interest was in the building of commercial transports for the airlines and accordingly Consolidated set about developing a version of the XPY-1 to meet the needs of this market. This seaplane would accommodate 22 passengers in quite luxurious surroundings, and would have a crew of three. This design was named the Commodore (CAC Model 16) and six were quickly sold to Tri-Motor in March 1929.

The airplane, based on the XPY-1, had a wingspan of 100 feet, a length of 61 feet 8 inches, and a gross weight of 17,600 pounds. It was powered by two Pratt & Whitney R-1860 Hornet B engines of 575 hp each, and had a maximum speed of 128 mph. Its range was 1,000 miles.

Tri-Motor was soon renamed the New York, Rio, Buenos Aires Lines (NYRBA) and provided both mail and passenger service between major cities in Central America, South America, and the United States, initially using Sikorsky S-38s. Through sales of Commodore airliners to NYRBA, Fleet was able to recover the investment it had made in the Navy XPY-1 prototype program and have a profit on the commercial design.

Consolidated went on to sell a total of 14 Commodores to NYRBA, which merged with Pan American Airways (PAA) in August 1930. The Commodores were used successfully for many years and PAA sold the last one in inventory in 1945.

Seaplane Programs

The Fleetster was conceived as a fast eight-passenger commercial monoplane that was generally similar to the Lockheed Vega. It was only semi successful as far as sales, but a small quantity (10) was sold to NYRBA about the same time as the Commodore, for use in South America. These were built in both a landplane configuration and the float seaplane configuration shown here. (Convair via SDASM)

As the XPY-1 program was nearing completion and the Commodore was getting underway, Major Fleet announced in March 1929 that Consolidated was going to build a high-performance, fast passenger (six to eight) and mail transport to be called the Fleetster (CAC Model 17). This was a high-wing monoplane generally resembling the earlier Lockheed Vega, although it would have an all-metal monocoque fuselage—the first in the United States—in lieu of the Vega's wood fuselage. It would also use the new National Advisory Committee for Aeronautics (NACA) low-drag engine cowl to improve the aerodynamic performance.

The Fleetster had a wingspan of 45 feet and was 31 feet 9 inches long. It was powered by a Pratt & Whitney R-1860-1 Hornet B engine rated at 575 hp, giving the Fleetster a top speed of 180 mph and a cruising speed of 150 mph, which was quite respectable at that time. Its gross weight was 5,600 pounds. A second version, the Model 20, had a parasol-wing configuration and initially an open cockpit that was later enclosed. The price of the Fleetster was $26,500.

The Fleetster first flew on 27 October 1929 and it was tested as a seaplane with twin floats soon thereafter in Miami, Florida. Tri-Motor Safety Airways was the first customer for this airplane and bought six Model 17 versions initially, and eventually four more, for mail and passengers on NYRBA's South American routes. Other customers included TWA (Model 20), the Army, and the Navy, which bought one as the XBY-1. Some of these airplanes later found their way to Russia and reportedly to the Spanish government. Fleet, however, did not consider the Fleetster a particularly successful program with sales totaling only 25 airplanes; barely 20 percent of the number of Lockheed Vegas sold.

After losing the production contract for the XPY-1, Consolidated continued to work on the patrol seaplane design and was rewarded with a production order for the P2Y-1 Ranger in 1931 after Martin had difficulty in producing its P3M. The Ranger incorporated many features developed for the Commodore including an enclosed cockpit. *(Convair via SDASM)*

Further design refinements were incorporated in the Ranger follow-on production that was awarded in December 1933. The P2Y-2 included the major change of moving the engines into the wing leading edge. This was a significant improvement and kits were ordered to convert some of the P2Y-1s to the -2 configuration. *(Convair via SDASM)*

Seaplane Programs

eanwhile, Consolidated, having recouped its XPY-1 development investment with the Commodore program, began working on improved versions of its original flying-boat design. This strategy proved successful and Consolidated received the go-ahead from the Navy for an advanced version of the Admiral, to be designated the XP2Y-1 (CAC Model 22) Ranger. Improvements included streamlining the Commodore's enclosed cockpit, larger engines with the new NACA cowlings, cleaned-up struts and bracing arrangements, and a redesigned tail structure. It also introduced a new lower-drag sesquiplane configuration to support the outboard floats. The Navy was so impressed by the proposal that it ordered 23 Rangers even before the 26 March 1932 first flight. The prototype was tested with a third engine strut mounted above the wing, but it was soon discarded as impractical. The first delivery to the Navy of the Ranger was made on 1 February 1933.

The last production P2Y-1 received modifications and additional improvements at the factory, including larger 750-hp Wright Cyclone engines that were raised to the wing leading edge and a longer-chord NACA engine cowl. This version was designated the XP2Y-2 and kits were fabricated to convert some of the P2Y-1s to the -2 version. Twenty-three more of these new configurations were ordered as the P2Y-3. In addition to the Navy, Consolidated's foreign sales of the Ranger included one to Colombia, one to Japan, and six to Argentina. Total production of the very successful P2Y program, including the Commodore, was 71 airplanes.

In the Navy's 1933 competition to replace its Ranger fleet, Consolidated's XP3Y-1 was the winner over the Douglas XP3D-1. The Consolidated configuration had many innovations including integral fuel tanks in the wing and retractable wingtip floats. In this rare but low-resolution photo you can see that the design, especially the fuselage and hull, had a strong legacy to its immediate predecessor, the P2Y. (Convair via SDASM)

In 1933 the Navy initiated a procurement to develop a follow-on replacement long-range patrol aircraft for its fleet of P2Y Rangers and various other types still in service. This competition resulted in two awards, with Consolidated and Douglas being the winners for a competitive fly-off. Consolidated received a contract for one experimental aircraft, the XP3Y-1, on 28 October 1933, and Douglas was similarly contracted for the XP3D-1.

The twin-engine XP3Y-1 (CAC Model 28) design drew heavily on the configuration of the P2Y albeit very modernized. The fuselage was generally similar to the P2Y's, but the tail surfaces faired in. A pedestal-mounted wing replaced the multitude of struts, and wingtip retractable floats were incorporated to improve the streamlining of the airplane. Integral-wing fuel tanks, first in the industry, were another of the weight-saving advancements used by this design. The two 825-hp Pratt & Whitney Twin Wasp engines were mounted in the wing leading edge in the same manner as on the P2Y-2.

The XP3Y-1 was built and assembled in Buffalo and shipped by rail to Norfolk, Virginia, for flight testing in the spring of 1935. The first flight took place on 21 March, about six weeks after the Douglas XP3D-1 flew, and testing of the XP3Y-1 went exceptionally well. Although both entries met the Navy's technical requirements, the Consolidated entry ended up as the winner of the competition based on price. The airplane program that was to be one of Consolidated's biggest success stories was now underway.

Consolidated's new patrol bomber was redesignated the PBY-1, later named the Catalina, and a 60 airplane production contract was awarded on 29 June 1935. These production aircraft were to be built in the newly completed plant in San Diego, California. The first San Diego-built PBY-1 was delivered to the Navy on 5 October 1936 only one year after the plant was dedicated. By this time additional orders for the PBY were on the books for a total of 176 more flying boats, including

There was considerable interest in an amphibious version of the PBY design and it had been promoted for several years. This low-resolution artist rendering shows one of the alternatives, a rather unique design, considered during amphibious capability studies prior to the adoption of the PBY-5A design. In this version the main landing gear retracted upward and outward into streamlined external pods supported on the two wing struts. (Convair via SDASM)

50 PBY-2s and 66 PBY-3s. As the war approached, even more orders were received from both the Navy and foreign governments.

Reuben Fleet had long maintained that the PBY should have amphibious capability but was challenged on that idea on the grounds of complexity and sacrifice of hull space that would be occupied by landing gear. One of the alternatives investigated was to have the main gear retract in pods mounted between the two main wing struts. The fuselage-stowed gear prevailed, however, and the Navy placed an order for 33 PBY-4s, the last of which was converted into the prototype of the amphibious version of the Catalina, the XPBY-5A. In December 1939, an additional 200 PBY-5s were ordered, of which 33 were the amphibian version. Because of this order the production level was such that a major factory expansion was required that essentially doubled the size of the San Diego production facility.

The PBY was the largest and most successful sea-plane program ever with a total of 3,308 being produced. This includes 2,160 built in San Diego and 235 in New Orleans by Consolidated, 155 by the Naval Aircraft Factory, 731 in Canada, and 27 in Russia. The last production Catalina, a PBY-6A, was built in New Orleans and delivered in September 1945, almost nine years after the first production delivery. The last Catalina, also a PBY-6A, was retired from Navy service on 1 January 1957, about 22 years after the first flight of the XP3Y-1.

Consolidated was very fortunate to have a viable production program in the 1930s, as this was a difficult time in the industry in general while the economy struggled to recover during the Depression years. As the Consolidated archives show, these times in the late 1930s also inspired a vigorous period of advanced design activities. Many design studies for a wide variety of aircraft were conducted and many concepts and variations, both military and commercial, were investigated.

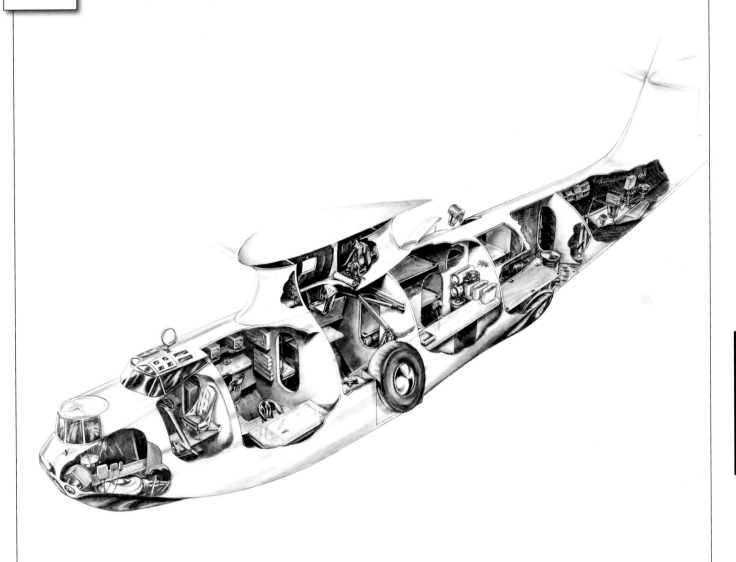

The PBY was proposed for aerial survey missions as this interior view shows. Again it is the basic PBY amphibious configuration outfitted with a full suite of aerial camera installations in both the fore and aft fuselage. It is not believed that any PBYs were built specifically for that mission although some may have been modified for aerial survey work. (Convair via SDASM)

From the archival documentation available, it appears very few variations of the basic PBY were offered, and indeed the wartime effort would probably have been aimed at quantity over variety. Two mission variations—search and rescue, and survey—primarily show differences in interior arrangement and in the equipment carried. Both of these missions may have actually materialized; for instance, in the Army's OA-10 or in some of the Navy aircraft. However, no factory changes were made on the type.

An advanced PBY configuration with further refined structure was shown in unidentified artwork that included a revised pedestal-mounted wing and elimination of the wing struts. It incorporated larger Pratt & Whitney R-2180 engines. It is not known if this was aimed at a specific mission or if it was a general design study.

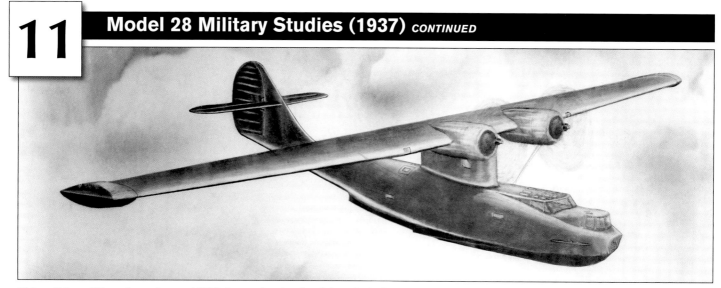

This artist rendition of an advanced PBY shows structural revisions including the elimination of the wing struts and a revised design of the wing pedestal. The engines were noted as the larger Pratt & Whitney R-2180s. It is not known if this was actually proposed to the Navy or if it was just an exploratory study. (Convair via SDASM)

The PBY fuselage was developed directly from the P2Y but because of the limited volume available, it was not very efficient as a transport or cargo airplane. Deep-hull versions were studied, this one possibly being a passenger transport because of the windows, although gun positions are also visible. The wing, engines, and tail are clearly from the PBY design. (Convair via SDASM)

The PBY fuselage with its heavy design legacy in the XPY and P2Y Navy flying boats, while quite workable for the military mission, was rather inefficient for a transport. The deep-hull version of a patrol bomber was investigated and clearly utilized the PBY wing and tail structure. Several deep-hull versions of the basic PBY airframe were studied for transport and cargo purposes as applied to the Navy's mission as well as the commercial market. It appears the Navy also looked at patrol bomber versions, although most of those studies centered on all-new flying-boat designs.

It has been suggested that an Allison-powered pusher version of the PBY arrangement was investigated, as was a four-engine version. No archival material was found to relate such configurations directly to the specific PBY design; however, similar arrangements were noted in various other studies during the late 1930s.

Seaplane Programs

Seaplane Programs

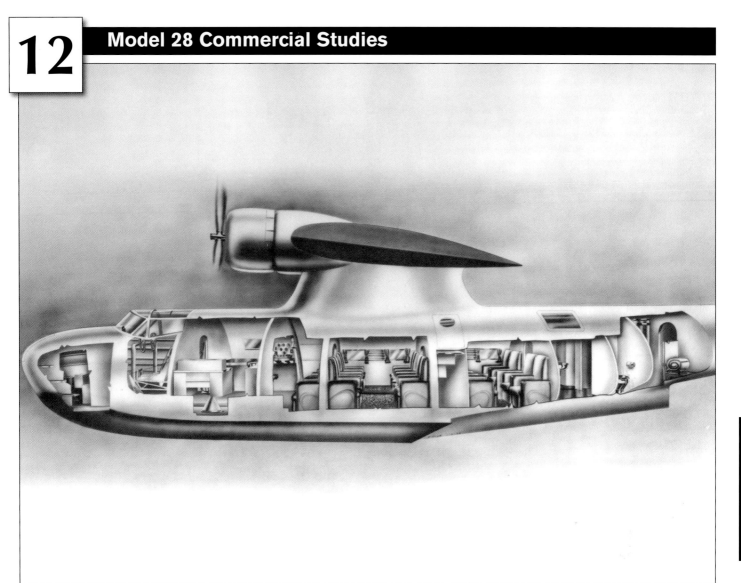

Here is an enhanced view of the 20-passenger Model 28 arrangement. The entrance door was located at the upper fuselage toward the rear of the wing pylon with stairs down to the passenger deck. The front compartment immediately behind the flight deck seated four and the other two compartments seated eight each. (Convair via SDASM)

Consolidated, with the reasonably successful experience with the Commodore version of the XPY-1 Admiral under its belt, looked to the PBY for possible commercial marketing potential. Its first opportunity came in response to an inquiry from Dr. Richard Archbold of the National Museum of Natural History. In response, Consolidated proposed and sold a commercial version of the PBY-1 (CAC Model 28-1) to Archbold for use in an expedition to New Guinea. The airplane was named the Guba (Indonesian term for wind storm). However, before the expedition was able to get underway Consolidated sold it to the Soviet Union.

The Russians had requested the airplane on an emergency basis to search for lost Russian aviators in the Arctic. Archbold subsequently ordered a second PBY, this time a commercial Model 28 (PBY-2), which was named Guba II. These Model 28s had been configured as an Air Yacht, a term originating with the earliest seaplane enterprises signifying an interior arrangement tailored for an individual buyer's needs, similar to today's business jets. In addition, their decks were beefed up to carry heavier loads, gun positions were replaced with cargo hatches, and other changes were made to that effect.

Many versions of a commercial PBY transport were studied, primarily with various passenger accommodations and interior arrangements. This flight view depicts one of the Model 28 commercial passenger versions and is recognizable mainly by the fuselage windows. (Convair via SDASM)

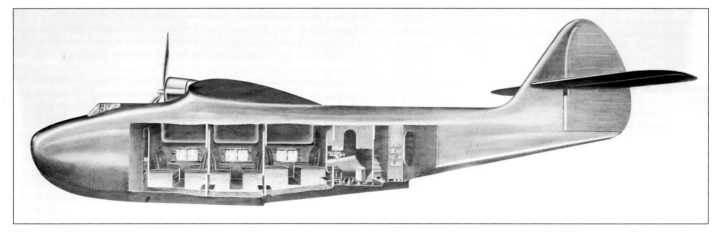

This deep-hull Model 28 type airplane was configured for 20 passengers and was not otherwise significantly different than the standard PBY passenger arrangement. The increased height, however, may be in recognition of the limitations of the PBY fuselage for this mission. This artwork included the name "Super Commodore" in the title, but it was the only time this notation was seen. (Convair via SDASM)

Other non-U.S. Navy sales of the PBY prior to World War II, in addition to the Gubas, included three additional PBY-1s to Russia, one to the American Export Airlines, and one to the British Air Ministry.

As the PBY production continued and stimulated by the sale of the Gubas and PBYs to American Export Airlines, and foreign companies, Consolidated looked to other commercial and non-Navy opportunities for this airframe design. Several passenger and cargo versions, both flying-boat and amphibian, were considered during this time, presumably being offered to the military as well as to commercial interests. These included passenger transport versions with interior arrangements with accommodations for 20 to a maximum of 34 passengers. Air Yacht versions were also configured with interior arrangements catering to potential individual buyers.

Deep-hull versions were also investigated, including a 20-passenger flying boat and an all-cargo configuration. The passenger version was named "Super Commodore" on some illustration artwork. A similar configuration with full-hull cargo accommodations was termed "Consolidated Air Express," but it is doubtful the name had any significance.

The advent of World War II and the attention required for the high-volume military airplane production generally halted any further consideration of civilian markets at this time.

Seaplane Programs

The four-engine XPB2Y-1 Coronado was only slightly larger in dimension than the PBY, by about 10 percent, but was about twice the gross weight. The XPB2Y-1 had serious directional stability problems discovered on its first flight, due to the single vertical stabilizer. An attempt at a modest fix with small additional vertical fins was unsuccessful and a complete empennage redesign was necessary. This solved the aerodynamic problem and resulted in the now-familiar Consolidated twin tail adapted for many of its airplane designs. (Convair via SDASM)

The Navy, after successfully fielding the Ranger patrol airplane and initiating the program for its replacement, the PBY, turned its attention to a larger and longer-range patrol bomber. The new requirement called for a seaplane with four engines and an ordnance payload of 12,000 pounds, or twice that of the PBY. A competition was held and was won by Consolidated and Sikorsky for a fly-off between the two. This award, in May 1936, was for $600,000, to which Consolidated also invested an additional $400,000. As was the Navy's habit at that time, Consolidated and the Navy were to cooperatively work together on the design of the airplane.

Consolidated's entry, the XPB2Y-1 Coronado (CAC Model 29), was a deep-hull flying boat powered by four 1,050-hp Pratt & Whitney XR-1830-78 Twin Wasp engines. It had a wingspan of 115 feet, was 79 feet 3 inches long, and had a gross weight of just under 50,000 pounds. Later production versions, however, saw an increase in gross weight to about 68,000 pounds. The PB2Y was designed to carry an ordnance payload of 12,000 pounds and had a crew of 10. Although the PB2Y had four engines and was significantly heavier, it was physically not that much larger than the PBY. The

PB2Y's wingspan was only 11 feet greater and its length was 13 feet more than that of the PBY.

The first flight took place on 17 December 1937 and it was immediately noted that there were serious problems with the airplane's directional stability, particularly in the power-off flight condition. After an early fix, of adding tail area in the form of small verticals outboard on the horizontal stabilizer, which did not provide an adequate solution, a complete redesign of the tail was accomplished. This resulted in the now-familiar Consolidated twin tail and was generally successful in solving those stability problems.

Production of the Coronado was not initiated until March 1939 because of the engineering problems and the commitment and priority of production of the PBY. The first order was for six PB2Y-2s which were used primarily for test purposes. These airplanes would have the new twin tail and a somewhat deeper hull, and would be equipped with upgraded 1,200-hp Twin Wasp 78 engines. Two of them were converted to XPB2Y-3 and XPB2Y-4. A total of 210 PB2Y-3s and -3Rs were built, a number of which were converted to PB2Y-5s. All in all, 217 PB2Ys were produced.

Seaplane Programs

A twin-engine PB2Y was offered in January 1940 that was to use the 2,000-hp Wright R-3350B engine. This provided a significantly lower gross weight and increase in performance. It appears that the remainder of the plane was basically unchanged. This twin-engine variant is similar to ones also offered later for the B-24 and B-32. (Convair via SDASM)

Available archival information suggests few advanced versions of the military PB2Y were investigated other than alternate interior arrangements appropriate to differing missions, such as for cargo and ambulance usage. This was probably due to the modest production and usage of this program and because neither the Navy nor Consolidated were particularly enamored with the airplane.

The single advanced version found (Design Study BM-13) was for a twin-engine configuration of the basic PB2Y patrol bomber airplane, whichwas offered in April 1940. In this proposed version two Wright R-3350B engines replaced the four Pratt & Whitney Twin Wasps. This powerplant change resulted in significant projected performance improvements. This version had a gross weight of 57,630 pounds, about 8,000 pounds less than the PB2Y-3. The resulting airplane would have a high speed of 240 mph, an increase of 23 mph, and a range increase of about 20 percent. The interior fuselage arrangements appeared to be essentially identical.

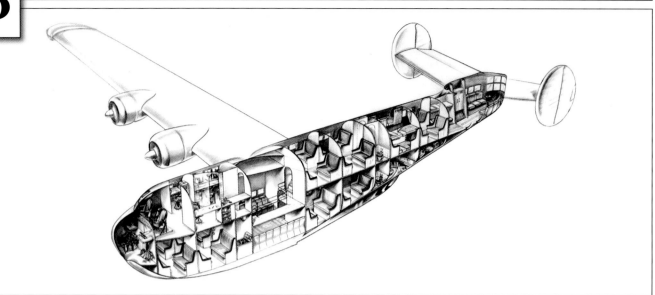

This commercial passenger version of the PB2Y features night sleeping accommodations for 38 passengers or alternately 60 passengers in day flight. The full-fuselage-width-lounge in this view is interesting in that it was also open through both upper and lower decks and included a walkway passage for the upper deck. The observation deck/lounge at the far aft of the fuselage under the tail appears to seat six. (Convair via SDASM)

The passenger versions of the PB2Y used the basic airframe virtually unchanged and only the interior was tailored for these configurations. Visually, only the absence of defensive armament and the addition of windows for each of the two decks and at the aft end reveal the transport version. The earliest studies were started in the latter part of 1937 and continued until 1939. Several variants were studied including both night and day version arrangements with passenger capacity varying from 12 to 60. (Convair via SDASM)

The stairway at the right of the lounge area led to the upper deck and the fuselage entrance door. The passageway at the rear provided access to the lower-deck seating compartments. The apparently unsecured casual chairs and table in this illustration probably would not survive a more thorough engineering review. (Convair via SDASM)

Several commercial passenger versions of the PB2Y (Design Study BC-2) were studied starting in November 1938 and lasted for about one year, with the last brochure dated November 1939. As with most of the commercial-market studies they tended to be overtaken by the events of the forthcoming war. The earliest configuration would accommodate 60 day passengers or 32 passengers for night flights with sleeping accommodations. Later versions were aimed at a more luxury accommodation for 12 to 18 passengers. In all cases the Model 29 airframe remained basically unchanged.

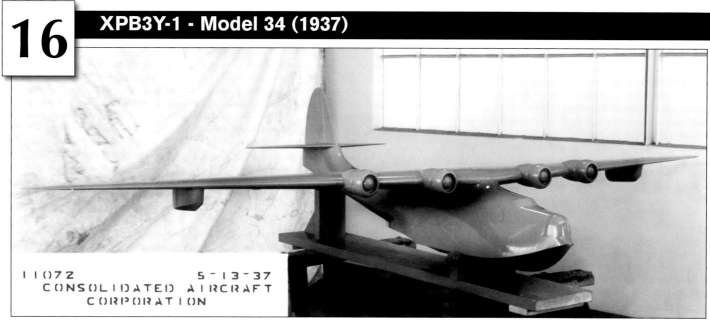

The XPB3Y-1 has an interesting history in that it may have been under contract for seven years before it was canceled in 1944, but it was never built. The earliest record was in May 1937 when it was apparently conceived in response to a Navy Requirement OS116-16. The wind tunnel model for the XPB3Y-1 shows the family resemblance to the PB2Y and it incorporates the PB2Y-1's initially unsuccessful single tail. (Convair via SDASM)

Seaplane Programs

The XPB3Y-1 program is somewhat of an enigma, at least from the very minimal and fragmented archival data that has survived. It was an interesting project in that there was considerable evolution of the design over a relatively long period; however, there was apparently never enough priority for the program to proceed. It is not clear if the lack of priority had to do with the customer, the requirements, or Consolidated's preoccupation with other current programs. There is little information concerning the circumstances of the original requirements being addressed, or the chronology. There is a reference to Navy requirement SD116-16 (May 1937) in the archives but no specifics were available. In any case this airplane appears to have been a step beyond the PB2Y with significantly enhanced capability requirements. Apparently, from the last available information on the airplane in August 1942, it was to be able to carry 20,000 pounds of bombs and to have a patrol range of 5,000 miles, a significant extension of PB2Y capability.

The initial configuration of this large four-engine patrol bomber was almost 75 percent larger in size and gross weight than the PB2Y. The design was fairly straightforward and had the appearance of being a "big brother" of the PB2Y. Several brochures in the archives

concerning the airplane offer insight into the XPB3Y-1's study evolution and are included here.

The program was initiated early in 1937 and continued into 1938 with a low level of activity. There is no evidence that this was a proposal in response to a competition or if it was initially a company-sponsored study. The first noted artwork was dated in January 1937 and it was logged with the designation XPB3Y-1, but it is not known if Consolidated was under actual contract at that time. Consolidated's use of designations lacked discipline on some occasions and may indicate more anticipation than fact. There is also information that the original studies of this airplane were designated the Model 30 in the company's system, but in all later configuration work this designation was changed to the Model 34.

The first brochure available, dated February 1938, was titled *Long Range Patrol Bomber (Plan II)* and included artwork captioned XPB3Y-1 (Modified), indicative of an alternate version and/or program approach. This configuration included many of the design characteristics of four-engine seaplane studies that had been conducted earlier, including the buried liquid-cooled engines and highly blended wing and fuselage. No information was found to indicate the

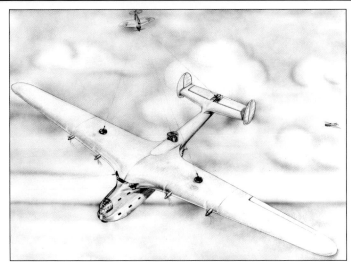

The first iteration of the XPB3Y-1 design was dated February 1938. The photo log included the title of "Flying Fortress" for this illustration of its defensive armament. It included the ventral retractable turret, the dorsal gun position, and a tail turret apparently able to fire upward and to some degree forward. Two gun positions on top of the wing midway between the engines were turrets, and were manned as seen on other designs. There were also gun positions on either side at the shoulder of the wing/fuselage interface. (Convair via SDASM)

The view from the Bridge "before the battle." Although the view is superlative, it is suspected that the Flight Commander's duties and the actual operability of such an arrangement had not been very well thought out. The navigator and the radio operator's positions were immediately in front of the Bridge. (Convair via SDASM)

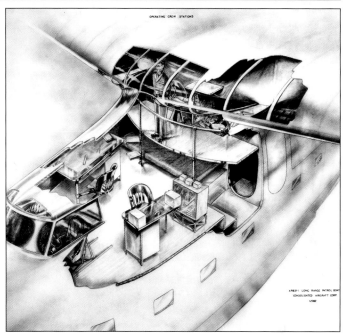

The Flight Commander's Bridge was incorporated in several studies located in the same wing/fuselage interface area and reflected the Navy shipboard manner of thinking. The flight deck arrangement is shown as quite spacious and included the pilot, copilot, navigator, and communications operator. This particular configuration is somewhat different in that it used a stepped cockpit and a different version of the Bridge. (Convair via SDASM)

design of the initial airplane (Plan I) but several differing concepts were also studied. One of the brochures that depicted the defensive armament of one version, interestingly, included artwork that the log identified by the name of the "Flying Fortress."

The final study effort on this patrol bomber was published in July and August 1942 when additional material was developed for the military configuration that was now visually very similar to an enlarged PB2Y. At this point it was powered by four Pratt & Whitney R-2800-18 two-stage, two-speed radial engines. It was to have a bomb load of twenty 1,000-pound bombs and defensive gun turrets in five locations.

The last data available indicates the contract was canceled in 1944 by mutual consent between Consolidated and the Navy after a prolonged period of inactivity without the program ever getting beyond the design phase.

Seaplane Programs

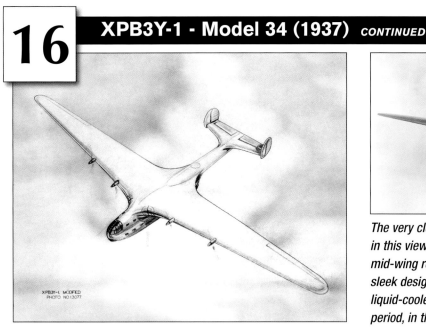

The XPB3Y-1 configuration in February 1938 exemplifies Consolidated's earlier design thinking for seaplanes including the rounded streamlined nose, the heavily blended wing/fuselage junction, the Model 31 short hull, and the buried liquid-cooled engines. This design had been modified and was labeled Plan II, but there is no information available as to the initial version. (Convair via SDASM)

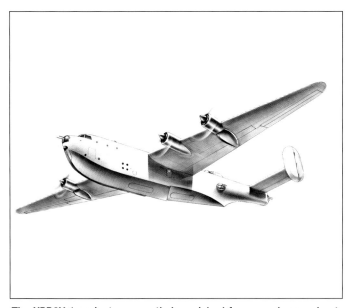

The XPB3Y-1 project apparently languished for several years due to the priorities of wartime production of the PBY and the PB2Y. A final revised version of this large patrol bomber appeared in July 1942. The short hull of 1938 had been replaced by a full-length fuselage hull and again had the appearance of a PB2Y big brother. It appears to have had bomb bay doors in the hull indicating conventional bomb racks in the fuselage. (Convair via SDASM)

The very clean aerodynamic design of this configuration is apparent in this view of the wind tunnel model. The characteristic twin tail, mid-wing retractable floats, and buried engines contribute to a very sleek design. The buried engines were again assumed to be the liquid-cooled Allison XV-3420s extensively used in studies of the period, in this case April 1938. (Convair via SDASM)

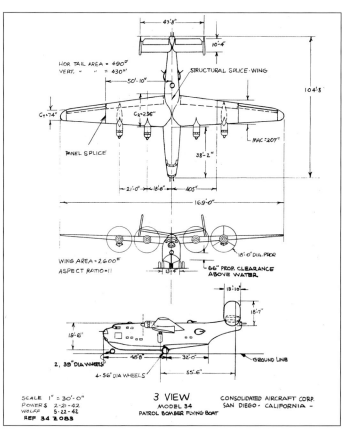

This iteration of the XPB3Y-1 had a wingspan of 169 feet, a length of 104 feet 8 inches, and a wing of 2,600 square feet. Four Pratt & Whitney R-2800 two-stage, two-speed 2,000-hp engines provided propulsion. The PB3Y-1 had a gross weight of 121,500 pounds, almost twice that of the PB2Y-1. It also had a range of 5,000 miles, a top speed of 237 mph, and was capable of carrying 20,000 pounds of bombs. (Convair via SDASM)

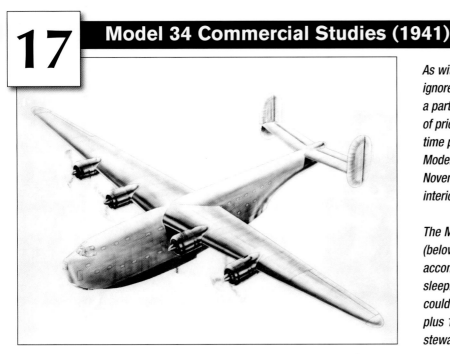

As with most of its seaplane projects, Consolidated did not ignore the transport and civilian application possibilities of a particular program. The XPB3Y-1 had suffered from lack of priority since 1938 but was revived in the 1941/1942 time period for both a military and a transport version. This Model 34 commercial transport design (left) studied in November 1941 used the patrol bomber airframe and an interior converted for passengers. (Convair via SDASM)

The Model 34 commercial transport accommodations (below) had a modern and familiar look. It could accommodate 77 day passengers or 41 night passengers sleeping in double berths on two decks. Alternatively it could accommodate 31 night passengers in single berths plus 10 chairs. The crew consisted of 14, including two stewards and one stewardess. (Convair via SDASM)

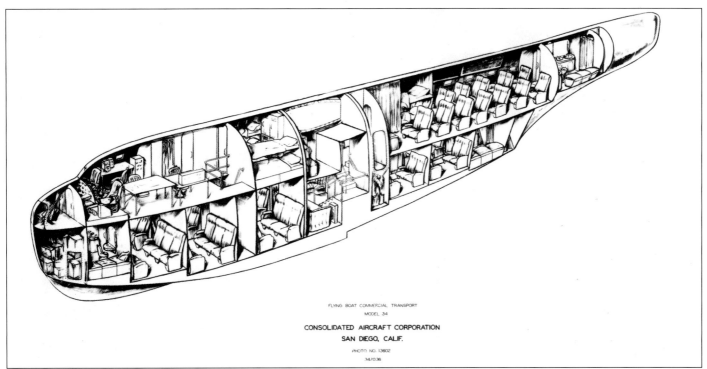

FLYING BOAT COMMERCIAL TRANSPORT
MODEL 34

CONSOLIDATED AIRCRAFT CORPORATION
SAN DIEGO, CALIF.

PHOTO NO. 13602
3470.36

Seaplane Programs

The archive records included a brochure on a commercial version of the Model 34 that was published in November 1941. This was several months before the last work on the XPB3Y-1 patrol bomber was conducted, or at least published. This Model 34 transport did utilize the final XPB3Y-1 airframe configuration. The design at this point had reverted to an airframe configuration visually much more like an enlarged Model 29 (PB2Y), and now employed radial engines. This civil version would carry 77 passengers in day accommodations or 41 in a night flight sleeper configuration. It also had a crew of 15 and had a range of 3,700 miles. The brochure included direct comparisons with the Boeing 314 and the Martin 130, so Consolidated apparently considered it a competitor, but the effort was cut short by World War II.

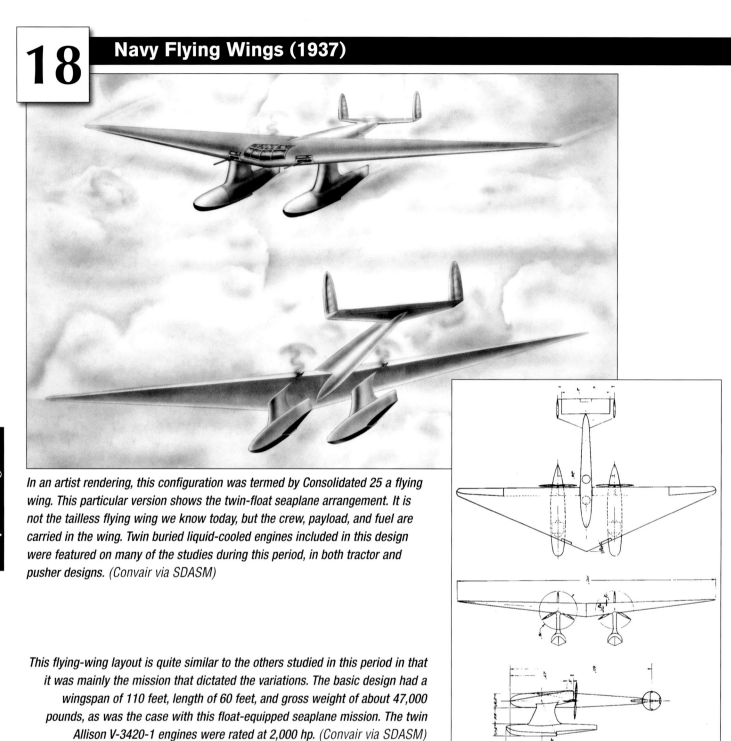

In an artist rendering, this configuration was termed by Consolidated 25 a flying wing. This particular version shows the twin-float seaplane arrangement. It is not the tailless flying wing we know today, but the crew, payload, and fuel are carried in the wing. Twin buried liquid-cooled engines included in this design were featured on many of the studies during this period, in both tractor and pusher designs. (Convair via SDASM)

This flying-wing layout is quite similar to the others studied in this period in that it was mainly the mission that dictated the variations. The basic design had a wingspan of 110 feet, length of 60 feet, and gross weight of about 47,000 pounds, as was the case with this float-equipped seaplane mission. The twin Allison V-3420-1 engines were rated at 2,000 hp. (Convair via SDASM)

Consolidated initiated studies of what was termed a flying wing configuration around 1937. This particular design arrangement was not a true tailless flying wing that we know today, but incorporated a conventional empennage supported by a minimal fuselage structure. The wing accommodated the engines, crew, armaments, passengers and/or cargo, and fuel. The configuration is somewhat similar to John K. Northrop's "all wing" Avion Model 1 test airplane first flown in 1929. Such competitive investigations of various prevalent ideas and configurations were quite widespread throughout the industry in this period of aviation's development.

Most of the consolidated flying-wing studies involved Army bombers and military or commercial transports, but

Seaplane Programs

NO. 12137   2-16-38
CONSOLIDATED AIRCRAFT
CORPORATION
WIND TUNNEL MODEL FLYING
WING - 110 FT. SPAN WITH
TWIN FLOATS.

This twin-tail, twin-float, wind tunnel model shows Consolidated's basic and original flying wing in a seaplane configuration, although the majority of the versions studied were landplanes. This model, in early February 1938, was tested in the Guggenheim Aeronautical Laboratory's wind tunnel at Caltech. (Convair via SDASM)

Seaplane Programs

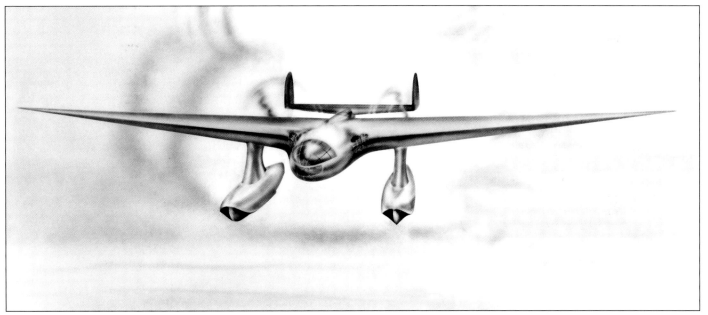

This artist rendering of the flying wing incorporates a more conventional nose for the flight crew. Note the windows at the wing root show gun positions. This was a popular feature in many of the studies at the time with designs that featured blended wings. This nose design may be a concession to the space limitations of the flying-wing configuration and may provide for more efficient space utilization. (Convair via SDASM)

several seaplane versions were also investigated. These twin-float seaplanes were versions of the basic two-engine pusher design that had a wingspan of 110 feet, a wing area of 1,650 square feet, and a gross weight of about 47,000 pounds. The powerplants were Allison XV-3420, 2,000-hp liquid-cooled engines that were buried in the wing. The aircraft had a capability to carry a military load of 4,000 pounds with a range of 4,000 miles and a top speed of 280 mph. It carried a crew of six and, as in the original configuration, located this flight crew in the leading edge of the wing. A second version had a more conventional extended forward fuselage for the crew. This flying-wing design approach apparently generated little interest and was abandoned by 1938.

Consolidated's Trans-Oceanic Seaplane study was in the time period of the Pan American World Airways' requirement for a new flying boat to replace the Sikorsky S-42s and the Martin 130s. The Boeing Model 314 won this competition. The Consolidated study was in the same size range and passenger capacity, but there is no evidence that this was actually submitted as a proposal; however, it is noted that the artwork did include PAA markings. (Convair via SDASM)

This configuration was for a 54-passenger flying boat that had a gross weight of 110,000 pounds. Artwork highlights a rather ugly nose and hull interface attempting to preserve a circular cross section of the fuselage and the requirements of the hull. It is an interesting viewpoint of using a small boat for passenger enplaning and deplaning rather than at a dock, possibly indicative of the consideration of undeveloped destinations. (Convair via SDASM)

By the late 1930s, Pan American Airways (PAA) had a near monopoly on the over water air routes. Initially PAA was using Sikorsky S-42B and Martin 130 flying boats, but in 1938 it placed an order with Boeing for six Model 314 Clippers. It is not known if this was the result of an industry-wide solicitation; in any case Consolidated had prepared a design in that time period to very similar requirements. Consolidated referred to this project as the Trans-Oceanic Flying Boat and the associated artwork shows this configuration in PAA markings. Most of the drawings and brochure material were dated July and August 1937.

Consolidated's Trans-Oceanic four-engine flying-boat's design was generally conventional except for the rather odd, and visually unattractive, rounded nose and its manner of fairing into the fuselage hull lines. The airplane was to have a gross weight of 110,000 pounds and would carry 54 passengers in stateroom-type accommodations. No other information or drawings are known to have survived and no further reference after 1937 was noted.

Seaplane Programs

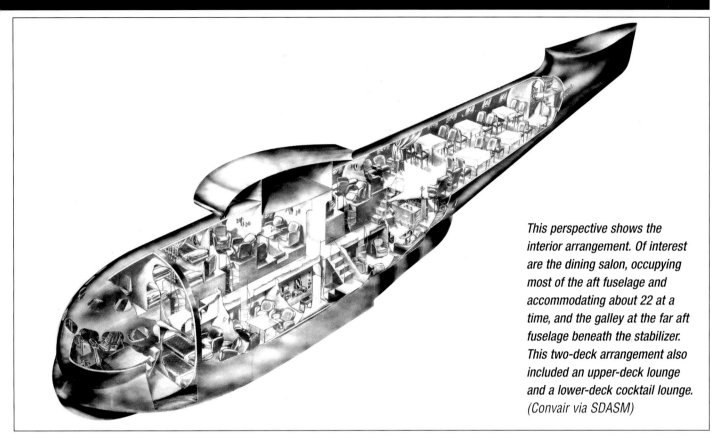

This perspective shows the interior arrangement. Of interest are the dining salon, occupying most of the aft fuselage and accommodating about 22 at a time, and the galley at the far aft fuselage beneath the stabilizer. This two-deck arrangement also included an upper-deck lounge and a lower-deck cocktail lounge. *(Convair via SDASM)*

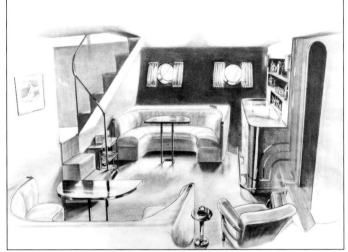

This is a view of the cocktail lounge area of the Trans-Oceanic Seaplane, a potentially crowded and high-traffic area. It is rather interesting to see the very early and somewhat naive vision of air travel as reflected by various components in this rendition. Some of these include a radio, freestanding ashtray, bar with a rail, apparently freely shelved glasses and bottles, plus the lack of seatbelts. All are familiar in the terrestrial domain but highly questionable in an aircraft. *(Convair via SDASM)*

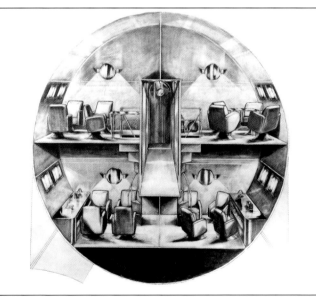

A fuselage cross-section of the seaplane shows passenger accommodations in the day section. As with the interiors of that time period, it is a rather luxurious layout, especially with regard to space, and certainly compared with the accommodations of contemporary air travel. Note the center upper-deck aisle way and the stairways to the lower deck. *(Convair via SDASM)*

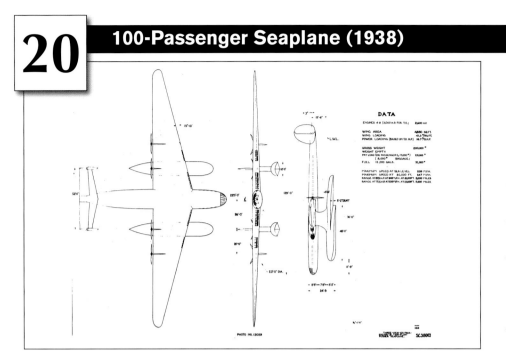

This 3-view of the float version of the 100-Passenger Seaplane highlights a clean design and the four buried liquid-cooled pusher engines rated at 2,600 hp (3,000 hp max). Note also the elliptical wide-body fuselage. It was a very large airplane with a 4,850-square-foot wing with a span of 220 feet. Its gross weight was 200,000 pounds. The float version turned out to be larger, heavier, and a little slower than the conventional version. (Convair via SDASM)

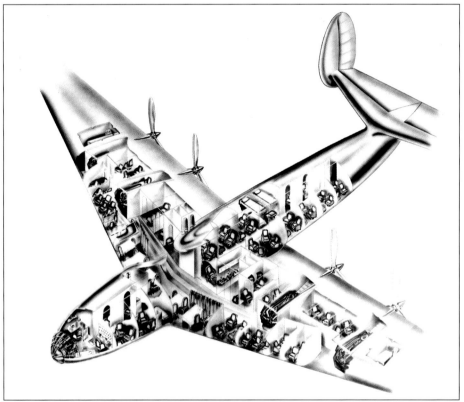

The 100-Passenger Seaplane preserves the individual cabin accommodations familiar from railroads and ocean liners. Interestingly, some passenger seats were arranged in "conversational" positions, as they were in many of the studies of the period. Dining compartments appear to be located in the wing leading edge and adjacent fuselage area. The idea of accommodating passengers and crew in the wing probably originated with the flying-wing studies carried out earlier. (Convair via SDASM)

A t the time that Consolidated announced it was studying the Trans-Oceanic Flying Boat, Reuben Fleet himself predicted that as soon as engines were available, giant flying boats carrying 100 passengers would be built to fly the over ocean routes.

It was said that the industry was quite surprised, however, when PAA, also looking to the future, requested proposals in December 1937 for a next-generation transport carrying 100 passengers for 5,000 miles. This airplane was envisioned to carry a 25,000-pound payload and cruise at 200 mph. The solicitation was requested in two parts, a preliminary response due in

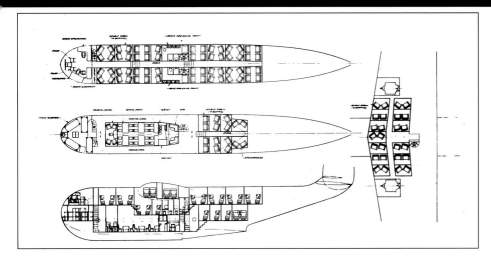

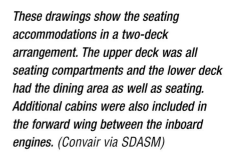

These drawings show the seating accommodations in a two-deck arrangement. The upper deck was all seating compartments and the lower deck had the dining area as well as seating. Additional cabins were also included in the forward wing between the inboard engines. (Convair via SDASM)

This more conventional configuration was developed in parallel with the floatplane approach. This design accommodates the passengers on two deck levels in the fuselage because the wing is available for only limited accommodations. It retained the buried liquid-cooled engines but reversed them to a tractor arrangement, and was somewhat smaller and about 30,000 pounds lighter than the float version. (Convair via SDASM)

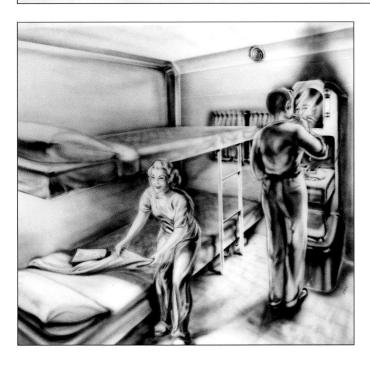

This luxurious and spacious compartment for two has fold-down bunk beds, as well as a toiletry unit. What happens to this unit during the day is not clear, nor is the disposition of the day seating provisions at night. (Convair via SDASM)

March 1938 and an additional proposal that could be awarded to selected contractors. Part Two would include additional work to include predesign and wind tunnel tests and would be funded at $35,000. Eight companies were solicited and it is believed that at a minimum Boeing, Douglas, Consolidated, Martin, and Sikorsky responded. It seems at least Part One of the solicitation was completed as artist renditions of the competitors' configurations appeared in industry media. This program apparently went dormant because of increasing tensions prior to the start of World War II. It is not known how Consolidated fared in this competition or even if selections were actually made.

Seaplane Programs

Archival documentation shows a Consolidated design that evolved in three stages. The initial configuration appeared in early January 1938 and was the most interesting but probably the least practical. This was a twin-float design and was probably based on several large airplane-float configurations that had been studied earlier. Four buried liquid-cooled engines powered it in a pusher arrangement. The resulting design was larger and heavier, and also slower than the more conventional designs that followed.

This approach was soon followed by an alternate more conventional configuration, and the new version incorporated a typical seaplane hull with the engines changed to a tractor propulsion arrangement.

Refinements continued and by submittal time in March, further changes were made, primarily in the shape of the hull. It retained the buried liquid-cooled tractor engines but the hull was shorter and had been truncated somewhat at the water line and incorporated a more upswept aft fuselage that supported the tail (similar to what would become the Model 31). It may have had a broadened fuselage, apparently to better accommodate the interior arrangement, and it also featured a more streamlined and shortened nose/cockpit/wing root arrangement.

It is interesting to note that these passenger seaplane interiors maintain the more luxury arrangements of staterooms in the tradition of the then-familiar steamship and railroad passenger accommodations. In addition, a fascinating and noteworthy feature in all three versions included passenger accommodations in the wing leading edge near the root of the fuselage, possibly inspired by work on the earlier Flying Wing configurations discussed above.

21   Prewar Two- and Three-Engine Flying Boats (1938)

The smaller of two Patrol Bomber (VPB) flying boats in this February 1938 study is slightly larger than the Model 31 and had the older pre-Davis-design wing but does have most of the other design characteristics of the Model 31 fuselage and twin tails. It also incorporated the twin buried liquid-cooled engines characteristic of the period. The study of this VPB and the larger one below it was conducted somewhat before the decision to build the Model 31. (Convair via SDASM)

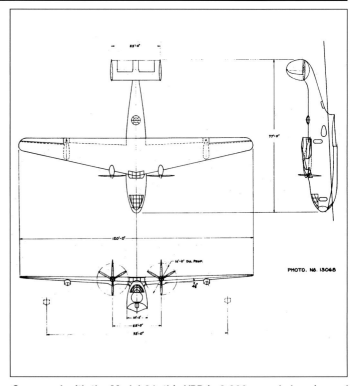

Several two-engine design studies were conducted prior to the decision to build the Model 31 and these studies led to its final design. The first of these, from brochures dated 15 February 1938, were two versions of a twin-engine VPB aircraft (both designated

Compared with the Model 31, this VPB is 2,000 pounds heavier and slightly larger, with a wingspan of 120 feet compared to 110 feet, and a length of 77 feet, compared to 73 feet. The maximum speed was estimated at 270 mph, and the range for a normal patrol mission was 4,750 statute miles. (Convair via SDASM)

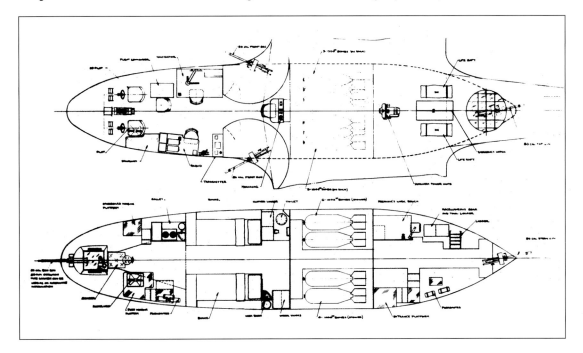

PHOTO NO. 8069

The flight deck of this VPB provides for a radio operator, navigator, and flight engineer just aft of the pilot's area. A unique aspect is the Flight Commander's Bridge located at the wing root and leading edge, quite reminiscent of a Navy surface-ship design. The first evidence of this design feature was in some of the PB3Y-1 designs. Note there were five gun positions, all manual or flexible mountings. (Convair via SDASM)

The forward shoulder gun positions are clearly shown in this plan view. The guns were .50-cal. and appear to have been manual flexible positions. The lower of the two decks included sleeping bunks, galley, and lavatory. The upper deck accommodated the flight crew and the defensive gun positions. (Convair via SDASM)

Seaplane Programs

BM-38), one with a wing area of 1,500 square feet and the other with 1,650 square feet. In both cases, buried Allison XV-3420-1 liquid-cooled engines powered them. This engine was basically a pair of V-1710 engines geared to a common prop that had a normal rating of 2,000 hp, and 2,300 hp maximum for takeoff. There was apparently considerable interest in this engine in the 1939 time period and it was actually scheduled for the Fisher P-75 fighter; however, the engine never completed development. It was believed to have flown on the XP-75 and been tested by General Motors on the Boeing XB-39. It

seemed to be a favored engine for many of the studies conducted by Consolidated, but quickly faded in view of its unavailability and the Navy's preference for air-cooled radial engines. Indeed, March 1938 artwork shows a very similar design of a VPB seaplane but with radial engines.

In both of these studies the fuselage and hull were clearly a predecessor of the Model 31 design with the shorter hull length at the waterline and the upturned rear fuselage. Both of these VPB configurations featured a rounded streamlined nose and forward fuselage and a heavy blending of the fuselage and the wing root. These

SeaPlane Programs

Consolidated investigated tri-motor concepts for landplanes and seaplanes, both for military and commercial purposes in this time period. This design in September 1938 was a Model 31-size patrol bomber with three of the Allison V-3420 liquid-cooled engines delivering 2,000 hp each. This seaplane had many of the design characteristics of the Model 31 and it appears to have had the Davis wing, judging by the aspect ratio and wing area. *(Convair via SDASM)*

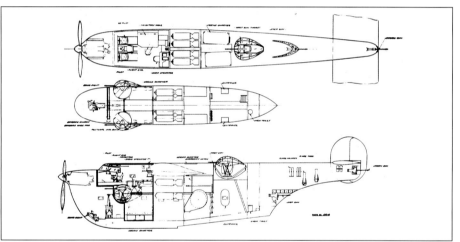

Inboard profile of the three-engine VPB shows five gun positions, two forward-fuselage side locations, a top and a ventral position, and a tail gun. All appear to be single .50-cal. flexible-mounted guns. Eight 1,000-pound bombs carried in the fuselage are moved through side doors for release, as was done on many Convair designs in the period. *(Convair via SDASM)*

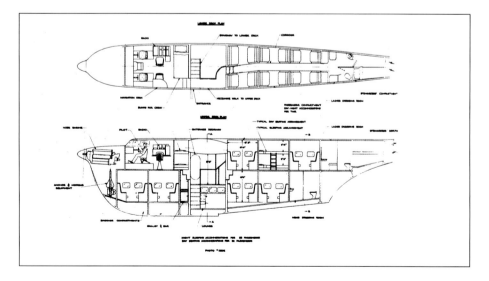

The interior of this three-engine passenger configuration is virtually identical to the commercial Model 31 versions except for differences due to the fuselage engine. The 52 day or 38 night passengers were accommodated on two decks in seven compartments. The entrance was on the upper deck directly under the wing and below what was a small lounge. *(Convair via SDASM)*

design characteristics were incorporated in several of Consolidated's studies in that period. These two seaplanes were somewhat larger than the Model 31. Both had 120-foot wingspans and the smaller had a normal gross weight of 52,000 pounds and the larger 55,000 pounds. The latter had an increased useful load capability of about 2,000 pounds but both are quoted as having the same normal patrol range of 4,750 statute miles (smi).

In September 1938, after the Model 31 program was underway, a study was conducted concerning three-engine versions of what was referred to as a modified Model 31. The basic airframe appears to be identical to the Model 31 but incorporated a nose engine as well as two engines buried in the wing, all liquid-cooled Allison XV-3420s. One version was a VPB (Design BM-6) and the other was a commercial pas-

senger transport. The three-engine VPB had the same gross weight as the Model 31 of 50,000 pounds and had a maximum speed of 338 mph. The normal patrol range with 1,000 pounds of bombs was 5,840 smi. Several tri-motor Army bomber versions were investigated in the same time period.

The commercial passenger seaplane variant of the tri-motor was configured for 52 day passengers or for 28 night flight sleeper accommodations and a crew of five, the same as the Model 31 commercial version. This three-engine design had an increased gross weight to 60,000 pounds versus 50,000 pounds for the Model 31, and had a top speed of 345 mph. The maximum range was 4,100 smi, at a cruising speed of 300 mph. This was apparently an exploratory study that emphasized speed for the passenger travel market.

22 Prewar Four-Engine Flying Boats (1938)

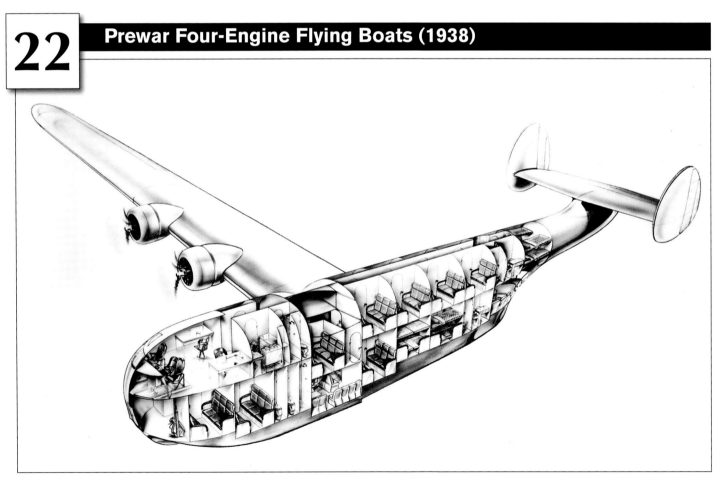

This four-engine commercial seaplane from September 1938 was configured to accommodate 38 passengers in a straightforward two-deck passenger arrangement. It seems similar to some Model 29 commercial versions although it does have the rounded nose of earlier designs. (Convair via SDASM)

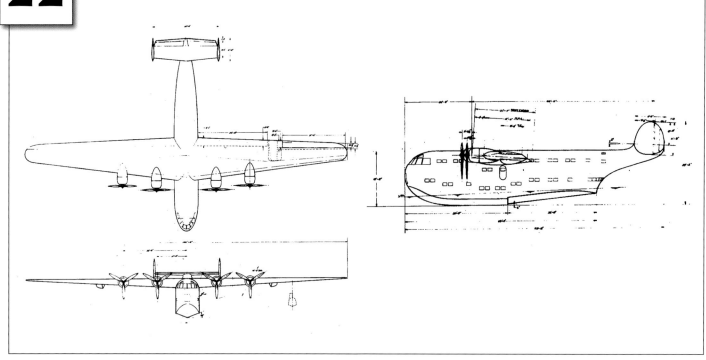

This commercial flying boat of September 1938 was fairly large with a design gross weight of 115,000 pounds and a wingspan of 162 feet. This is much larger than the Model 29 commercial versions that followed, with similar passenger capacity. This configuration had a shorter more rounded nose and earlier tail-surface shape but was otherwise undistinguished. (Convair via SDASM)

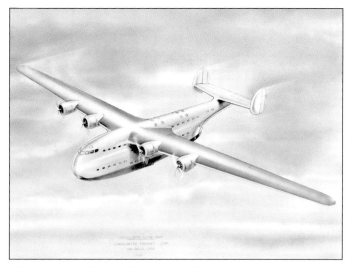

Although this artist rendition of an inflight view of the airplane was included in the brochure for a 38-passenger commercial transport, the artwork was dated six months earlier and gives the appearance of a larger airplane. It also seems to have a more refined appearance so it may be a different design than the 38-passenger seaplane. One of the big visual differences is the large number of passenger windows and the more streamlined nose. (Convair via SDASM)

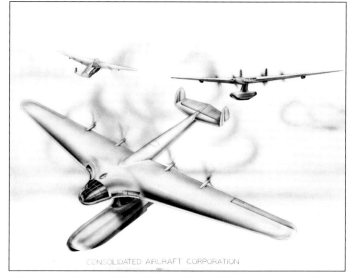

This very unique configuration, believed to be a patrol aircraft from February 1938, features four liquid-cooled pusher engines and a large single float. It also had a very short forward fuselage (rounded and highly glassed-in) and a very blended wing/fuselage junction. Also shown are crew positions in the wing root. This design may have been a further development of the flying wing designs studied earlier. (Convair via SDASM)

Very little information is available with regard to this rail-tracked takeoff device. It shows one of the earlier four-engine passenger seaplanes being launched from a powered high-speed rail vehicle at a land base. It is assumed to allow much higher gross weight than a normal water takeoff. The motivation and operational procedure or utility for such a system was not available and only the artwork survives. (Convair via SDASM)

The time period of late 1937 into 1939 was very prolific in terms of the number of design studies and the evolution of design trends for Consolidated. This is shown in the designs of four-engine flying boats and also can be observed in other types of seaplanes that had been investigated in this period. It was evident that there was a management emphasis on predesign activities to develop market areas to gain new business. Undoubtedly, increasing world tensions contributed to these activities, particularly on the military side, but even so, the possibilities of the rapidly developing commercial market were not lost on management and the designers.

Several of the design studies were not directly associated with the ongoing seaplane programs such as the PBY and the PB2Y. The first is a commercial 38-passenger four-engine flying boat studied in June 1938 that seems

fairly large for only 38 passengers. It is a bit of a puzzle because the brochure has artwork of the airplane in flight that has all the appearance of a much larger airplane.

A more radical design from slightly earlier in February 1938 was a large four-engine single-float, apparently military seaplane. This design incorporated the liquid-cooled engines in a pusher arrangement. It also had the highly aerodynamic wing/fuselage junction and the rounded nose seen on many designs. It also had the wing root crew positions.

In August 1938, Consolidated briefly examined the possibility of launching large seaplanes from a land-based rail-tracked takeoff device. The idea was apparently to be able to greatly increase the takeoff gross weight and attendent range increase, but it is not clear how it would be operationally employed.

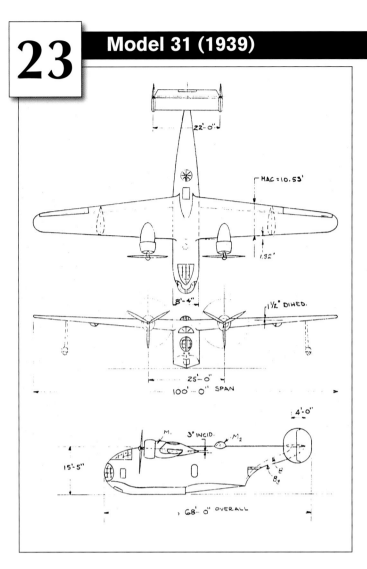

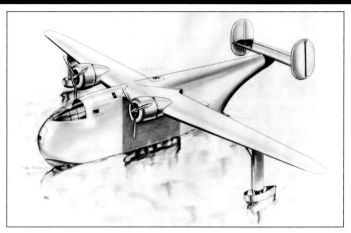

Artist rendition shows the Model X on the water and the mid-wing retractable floats are clearly visible. This view apparently illustrates the initial prototype airplane design that was intended to be built since it does not have gun positions. The rounded nose soon gave way to the familiar stepped cockpit that the eventual design employed. (Convair via SDASM)

The early work on the twin-engine Model 31 was conducted under the designations Model X and 31X. The initial configuration had a rounded nose that was prevalent on many of the design studies at the time. The earliest configurations, before the Davis-design wing was adopted, had a 100-foot wingspan and a 68-foot length. At least three gun positions are visible in the drawing. (Convair via SDASM)

Seaplane Programs

The PBY program was in full production and the PB2Y had just flown for the first time. In that same time period Martin was working with the Navy on the development of a new advanced patrol bomber, the PBM-1 Mariner (Martin Model 162) with two R-2600 engines. Reuben Fleet and his staff had decided to undertake a new clean-sheet design of a similar high-performance twin-engine seaplane using the new Wright R-3350 Cyclone engines that would be a significant advance over the PBY and the PB2Y. Further, it could be carried out without what was then felt was customer interference. Fleet considered that he had a less-than-satisfactory experience resulting from collaboration with the Navy on the PB2Y design. There was apparently considerable conflict with regard to philosophy and design details that resulted in neither party being pleased.

The predesign of what was to be designated the Model 31 was based on the ongoing studies of twin-engine patrol bombers discussed previously. Initially it was known as Model X and then 31X and the first designs had a rounded nose that was soon changed to the final stepped cockpit. Engineering plans were firmed up and Consolidated then committed to invest the necessary funds, about $1 million, to accomplish the detailed design work and build a prototype airplane as a company-sponsored project, the Model 31.

At about this same time, an aerodynamic engineer, David R. Davis, from outside the company, approached Fleet seeking to sell his new analytical approach to wing airfoil design. Airfoils had generally been arrived at by trial-and-error wind tunnel testing. Fleet was skeptical but did reach agreement with Davis in February 1938 to conduct tests at the Caltech (GALCIT) wind tunnel. The tunnel tests turned out to be very impressive and showed a significant increase in airfoil efficiency. Since

The prototype twin-engine Model 31 is shown with its civilian tail number as a company-funded non-military program. The massive deep-hull design quickly evoked the nickname "The Pregnant Guppy," although its official name under the later aborted P4Y-1 Navy program was the Corregidor. Note the incorporation of a fully retractable beaching gear, obviating the need for separate support equipment and the necessity to attach and remove such gear. (Convair via SDASM)

the Model 31 design was just getting underway, it appeared to be a good candidate and an opportune time to incorporate the Davis wing design and determine if the predicted performance improvements would be sustained in full-scale application.

Fabrication of the Model 31 began in July 1938 and the prototype was rolled out 10 months later. It had a very deep hull, about 22 feet, and an upswept tail supporting twin vertical stabilizers similar to those adopted for the XPB2Y-1. This resulted in a rather unattractive appearance because of the fuselage shape and quickly gained the nickname of Pregnant Guppy. The high-aspect-ratio (A/R) Davis wing had a span of 110 feet but omitted the familiar retractable wingtip floats of the PBY

and the PB2Y. These were not incorporated in view of the potential use of the wing design for landplanes. The design did include integral and fully retractable beaching gear. The powerplants for the Model 31 were the Wright R-3350 Cyclone engines rated at 2,000 hp, and it was the first airplane to use that engine.

The Model 31 was first tested on 5 May 1939. The flight tests were very successful and the optimistic predictions were vindicated with performance being improved by as much as 20 percent. The top speed of the Model 31 was 250 mph and the range was 3,500 miles. Later, after flight testing, the Model 31 airplane was acquired by the Navy and was modified to the XP4Y-1 configuration.

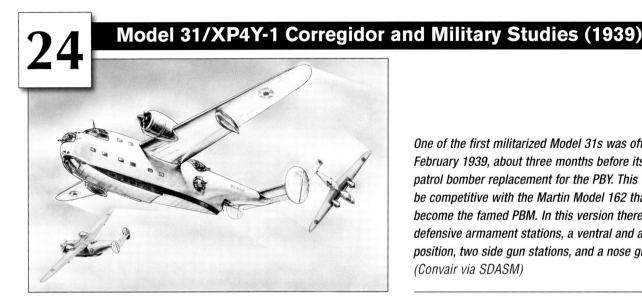

The private-venture Model 31 turned out to have significantly enhanced performance, but even so the Navy remained relatively uninterested because, in Fleet's view, they had not participated in the airplane's development. In addition, the Navy was involved during this period with the initiation of the Martin PBM seaplane of similar capability. Consolidated offered a medium-range patrol bomber version of the Model 31 as early as February 1939. That brochure described a basic Model 31 militarized airframe with the two Wright R-3350 radial engines that provided a maximum speed of 261 mph and a range of 3,800 smi. It had a gross weight of 47,216 pounds and carried four 1,000-pound bombs.

A rather unique study was conducted in April 1939 for an autogyro-configured Model 31. It featured a retractable single-blade counterbalanced rotor. It is assumed the idea was to take advantage of the very short takeoff and landing distances of the autogyro. This airplane configuration also permitted greatly reduced speeds and an inherent resistance to stalling. It is not known what use or mission was being proposed but anti-submarine warfare (ASW) may have been a possibility.

In November 1939, brochures were published with the results of Consolidated's study response to Navy Specification 116-23. It was initially thought it might have been associated with the Navy's PBY Catalina replacement program but some of the results indicate it was an exploratory design study. The detail requirements of the Specification are not known, but the response included two versions of the Model 31 patrol bomber. The first was a slightly enlarged version of the Model 31 (Design BM-7) and the other was a basic

original-size Model 31 (Design BM-8). Several engine options were considered including the Pratt & Whitney R-2800, the Wright R-3350, and in the case of BM-7, the Pratt & Whitney H-3130-AG2.

The unique aspect of the study is that catapult versions of these patrol bombers were featured along with a non-catapult design. The objective of the catapult launchings appears to have been to provide for a greatly increased gross weight and attendant range capability. For example the BM-7-design gross weight was increased from 55,554 to 89,092 pounds for the catapult launched version and the range was increased from 4,000 to 9,300 smi.

Later in 1940 the basic Model 31 was again offered to the Navy, this time with alternate engines that had been investigated for applicability and their attendant performance capabilities determined. This sales effort of the Model 31 was probably related to the impending Navy order for the PBM-1 (Model 162) that took place in the fall of 1940.

The Navy finally did acquire the prototype Model 31 in April 1942 as the XP4Y-1, and conducted armament testing. The Navy then decided to put the Model 31 into production as the P4Y-1 Corregidor and a 200-airplane order for what was believed to be an ASW role was placed in October. This activity was to take place at a New Orleans plant newly assigned to Consolidated for PBY production. The P4Y-1 was to have a modified bow and a raised tail. The plans for production were short lived and never got underway. The program was canceled when the decision was made to allocate the entire production of the Wright engines to the newly initiated B-29 program.

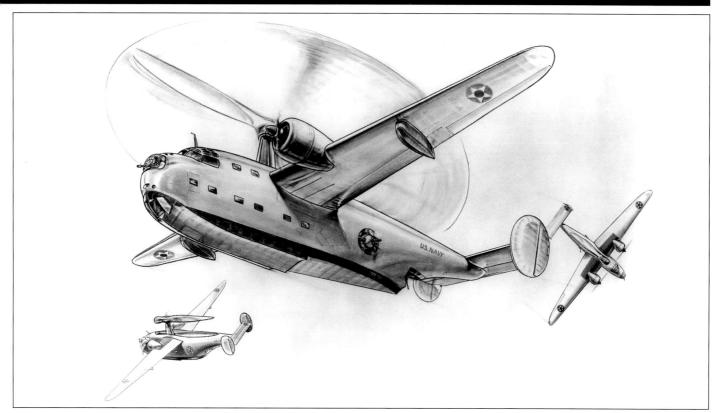

A startlingly unique configuration of a Model 31 patrol bomber variant is depicted in this view. It is designed as an autogiro version, apparently to take advantage of the capability of this type of aircraft for very short takeoffs and landings, and very slow flight. The rotor has a single counterbalanced blade and is retractable into the top of the fuselage for normal flight. (Convair via SDASM)

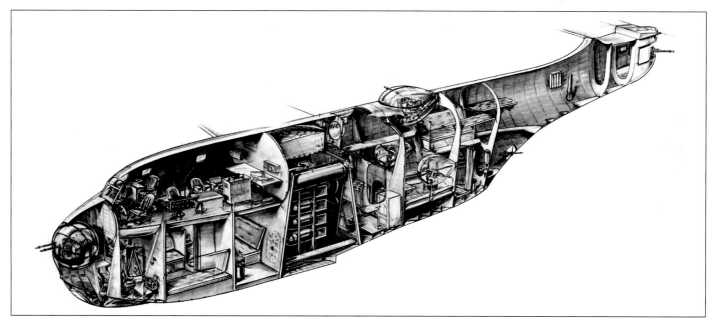

In November 1939, Consolidated submitted brochures, apparently responding to Navy Requirement Specification 116-23. The submittals were for two patrol bombers, one being the Model 31 (Design BM-8) and the other a somewhat enlarged version of the Model 31 (Design BM-7). The BM-8 version was quite comparable to the prior proposals per the artist rendition shown here. (Convair via SDASM)

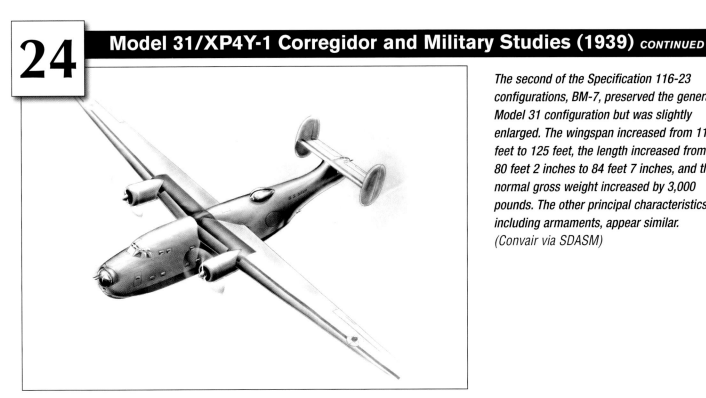

The second of the Specification 116-23 configurations, BM-7, preserved the general Model 31 configuration but was slightly enlarged. The wingspan increased from 110 feet to 125 feet, the length increased from 80 feet 2 inches to 84 feet 7 inches, and the normal gross weight increased by 3,000 pounds. The other principal characteristics, including armaments, appear similar. (Convair via SDASM)

The Model 31 was sold to the U.S. Navy in April 1942 and underwent flight testing. Some modifications were made to convert this airplane to the prototype XP4Y-1, but armament was not added for this flight testing. The Navy decided to produce the P4Y at the New Orleans plant as an ASW weapon but the project was canceled well before any actual airplanes were built. (Convair via SDASM)

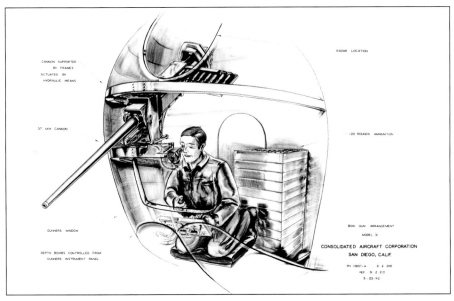

An early rendition of the bow cannon installation on the P4Y-1 depicts a kneeling gunner controlling the depth-bomb armament using the gunner's instrument panel. The 37mm cannon ammunition magazines were stored behind the gunner and were apparently hand loaded into the cannon itself, which would seem somewhat awkward in actual operation. In all likelihood the cannon was for use in the ASW role. (Convair via SDASM)

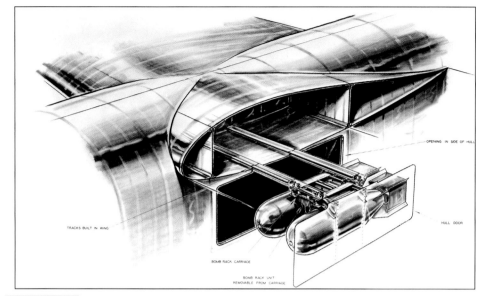

Bombs in the P4Y-1 were carried in the fuselage just below the wing and were mounted on individual racks that fit on a bomb-rack carriage. This carriage moved on tracks mounted in the wing and moved the bombs through a fuselage door into the airstream for release. (Convair via SDASM)

25 Model 31 Commercial Studies (1938)

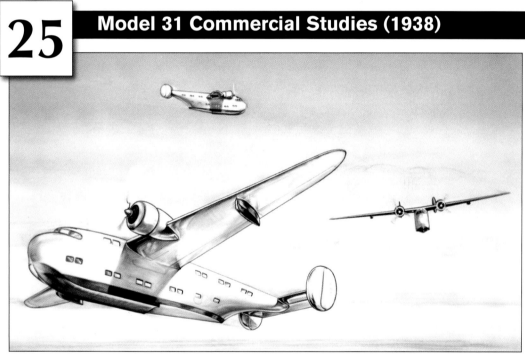

Mindful of the commercial potential, several Air Yacht versions and passenger configurations of the Model 31 were investigated. The same basic Model 31 airframe was used, and interiors were tailored for specific applications. Externally, other than the obvious deletion of armament, the only visible difference from the military version is the addition of windows for both upper and lower decks. (Convair via SDASM)

The first Model 31 design study for the commercial market was an Air Yacht configured for Dr. Richard Archbold of the New York Natural History Museum. Archbold had previously purchased a PBY, named Guba II, which he used for a New Guinea expedition. It is not known if the study was conducted at Consolidated's initiative or if was at Archbold's request. Additional Air Yacht configurations were studied that were oriented to a more executive use comparable to the widespread use of the business-configured aircraft of today.

Several commercial passenger configurations were also investigated in 1938 and incorporated generally conventional passenger accommodations. These interiors would accommodate up to 32 passengers for night flights or 60 passengers for day flights. One version carrying 28 passengers had a gross weight of 50,000 pounds and a range of between 2,000 and 3,000 smi, depending on ground rules.

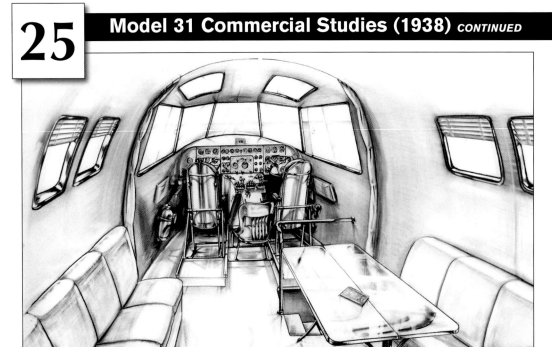

Lounge area of a proposed Archbold Air Yacht of July 1938 was located on the upper level immediately behind the flight deck, and included seating for up to 10 people, together with a large, table work area. The lounge area was open to the cockpit and stairway access to the lower deck was to the forward right end of this area. (Convair via SDASM)

Seaplane Programs

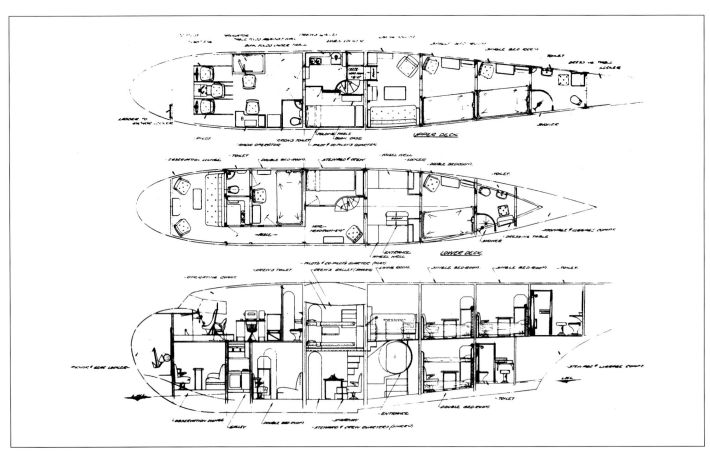

The generalized Air Yacht version is seen in this inboard profile drawing somewhat later in December 1938 and reflects a more "executive business" interior arrangement than that previously configured for the Archbold requirements. The nose was stepped in the manner of the Model 31. The flight deck was spacious, apparently accommodating a navigator and a communications crewmember in addition to the pilot and copilot. (Convair via SDASM)

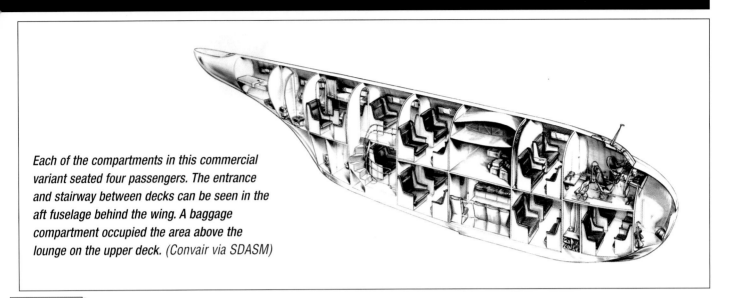

Each of the compartments in this commercial variant seated four passengers. The entrance and stairway between decks can be seen in the aft fuselage behind the wing. A baggage compartment occupied the area above the lounge on the upper deck. *(Convair via SDASM)*

26   PJY Utility Seaplane (1940)

Seaplane Programs

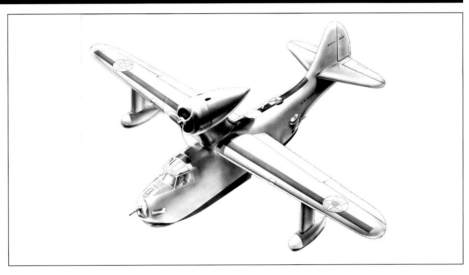

The BM-12 design, termed PJY-1 by Consolidated, is a small conventional high-wing utility seaplane with a single engine mounted on a pedestal above the fuselage. Little information on this proposal survived and the engine type is unknown. As may be seen, this version was armed with three gun positions. *(Convair via SDASM)*

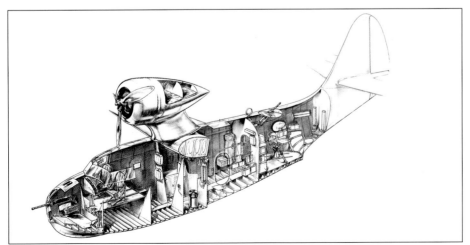

The interior of the BM-12 would indicate a crew of three, or possibly four, individuals. Defensive gun positions included a .50-cal. gun in the nose and a .30-cal. gun at a dorsal aft fuselage location. There was also a ventral gun position covering part of the tail area where apparently the .30-cal. gun could be moved if needed. Also note that fuel was carried in the aft portion of the engine nacelle. *(Convair via SDASM)*

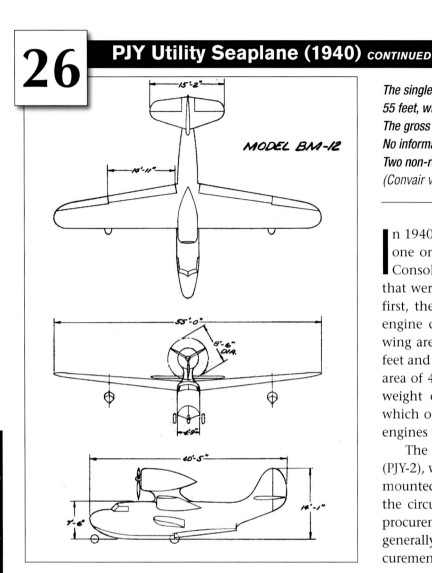

MODEL BM-12

The single-engine utility configuration had a wingspan of 55 feet, wing area of 375 square feet, and length of 40 feet 5 inches. The gross weight of this plane was 8,358 pounds.
No information is available as to the engine used in this proposal.
Two non-retractable wing floats stabilized the seaplane in the water. (Convair via SDASM)

In 1940 the Navy issued a specification (SD 116-28) for one or more aircraft types to fulfill a utility mission. Consolidated responded with several configurations that were referred to, at least internally, as the PJY. The first, the PJY-1 submitted 4 January 1940, was a twin-engine design in two versions differing principally in wing area. Design BM-9 had a wing area of 375 square feet and a span of 55 feet, and design BM-11 had a wing area of 450 square feet and a span of 60 feet. The gross weight of one of these two was 10,718 pounds but which one was not identified. It is also unknown what engines were to be used.

The second design dated 16 June 1940, the BM-12 (PJY-2), was a single-engine configuration with its engine mounted on a pedestal above the wing. It is not known the circumstances of this study or an associated Navy procurement and competition. It does seem, however, to generally coincide with the timing of the Navy procurement of the Grumman J4F-1 (Widgeon G44).

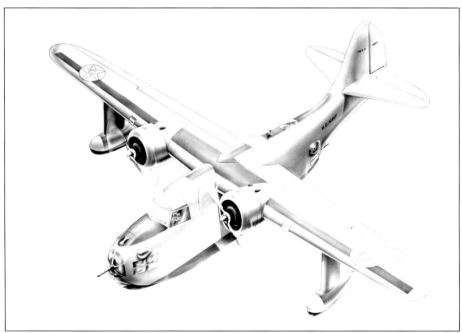

Design BM-11, also a utility seaplane (also termed in some places the PJY-1) is a straightforward twin-engine conventional seaplane. The fuselage appears similar to the single-engine version and seems to retain the identical three gun positions. (Convair via SDASM)

Seaplane Programs

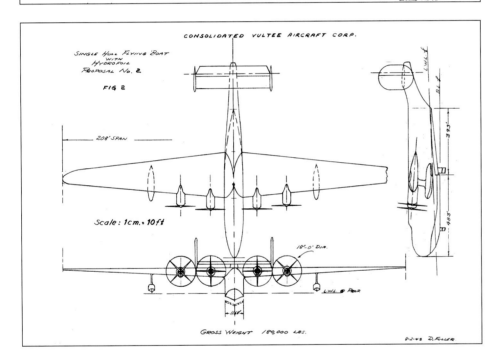

This study was conducted to determine the relative merits of a series of seaplane design approaches. The conventional single-hull, twin-tail, four-engine seaplane served as the reference comparison for other configurations considered. It was sized for a 180,000-pound gross weight, as were the other candidates in this study. It had a 208-foot wingspan and was powered by four Pratt & Whitney X Wasp Major engines. (Convair via SDASM)

The first conventional design has been modified in Design No. 2 with the addition of a high-aspect-ratio main hydrofoil and a smaller trimming foil on the V-bottom hull. Tests with a 1/8-scale XPB4Y-1 showed very good spray characteristics and the potential for a 20-percent beam reduction. (Convair via SDASM)

Ernest Stout, who oversaw the hydrodynamic research at Convair, was a staunch supporter of the seaplane programs and was the initiator of many concepts originated in the Hydrodynamics Department. One of the first studies conducted by Stout was dated August 1943. It was a wide-ranging generalized study aimed at improving seaplanes to be more competitive with landplane performance. Seaplanes tended to have inherently poorer performance than landplanes because of the hydrodynamic requirements of deep hulls and for lateral water stability. This particular study investigated a variety of innovative concepts including the relatively new technology of hydrofoils, increasing length-to-beam ratios (L/b), the use of canards, twin hulls, and the so-called Ventnor hull design. This latter design involved a tunnel in the middle of the hull to contain and channelize the spray to a high-velocity jet in the tunnel and minimize the spray external to the tunnel. This flying-boat-hull concept was believed to have been developed by Kaiser-Gar Wood and was tested by NACA.

Seaplane Programs

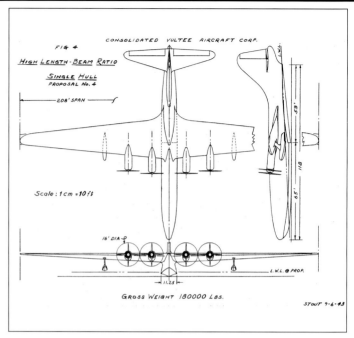

Design No. 3 employs the Ventnor type of hull that had a "tunnel" through its center. This hull was investigated by NACA and was thought to have very clean spray characteristics and low resistance that would reduce the hull beam by 20 percent and would allow a shallower hull. A disadvantage might have been a severe wetting of the tail due to exiting of the channelized high-velocity jet from the tunnel. *(Convair via SDASM)*

This configuration, Design No. 4, examined length-to-beam (L/b) ratio effects. As the L/b was increased it resulted in long forebodies, as the wing must remain over the hull step. It was felt that if the balance problem could be solved the high-L/b hulls would run very cleanly. *(Convair via SDASM)*

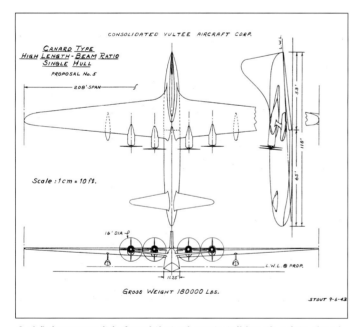

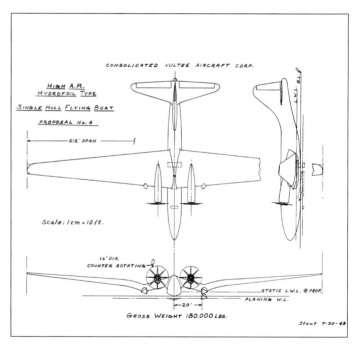

As L/b increases it is found that adequate tail length exists ahead of the center of gravity and the tail plane can be moved forward in a canard arrangement, as in Design No. 5. This also eliminated a disadvantage of high-L/b designs as they caused a large amount of water to be thrown over the tail. *(Convair via SDASM)*

Design No. 6 incorporated high-aspect-ratio hydrofoils and an unconventional gull wing aimed at reducing frontal area. Retractable floats were incorporated into the gull wing for on-water stability. *(Convair via SDASM)*

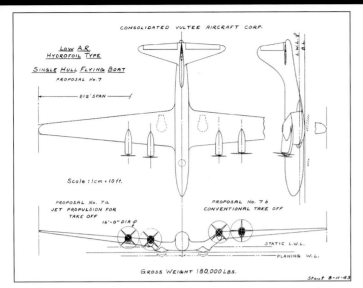

It was proposed to use low-aspect-ratio hydrofoils in Design No. 7 to overcome certain instability problems with the high-A/R foils at "hump speed." The configuration on the front view, right side (7b) considered conventional piston-engine takeoff where the propellers must remain clear. On the left, jet propulsion was used and the piston engines were not running at takeoff. *(Convair via SDASM)*

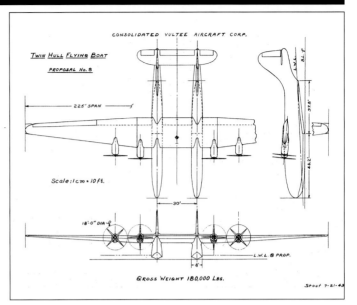

Twin-hull designs were thought to have considerable promise for using high-L/b fuselages and low "hump resistance" in Design No. 8. The use of the wing center section for crew and cargo was also pointed out. This twin-hull boat would use four Pratt & Whitney R-4360 Wasp Major engines of 3,250 hp each. *(Convair via SDASM)*

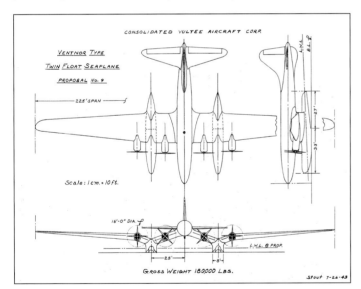

Design No. 9 was a twin-float version utilizing the Ventnor hull configuration and the unconventional gull wing. Because of the favorable manner of the spray from this design the propellers did not require the clearance ordinarily encountered. *(Convair via SDASM)*

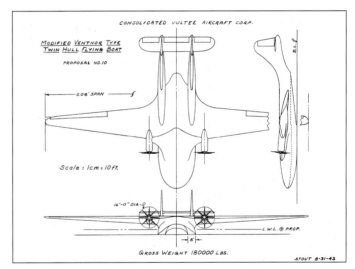

This modified Ventnor configuration in Design No. 10 envisions using the space between the hulls to channelize spray and reduce the beams of the two hulls. The planing surfaces were wide enough to provide static lateral stability thus eliminating the wingtip floats. *(Convair via SDASM)*

The conclusion from the study was that, in order to improve performance, it would be necessary to develop new innovations in hull design as a prime area for investigation. The ideas put forth in this study required much more research to substantiate the expected improvements.

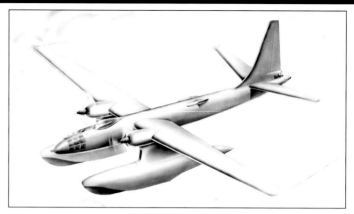

The first of the three compound-propulsion design configurations, from a patrol reconnaissance study in December 1944, is a quite conventional seaplane with two R-4360 engines plus two GE TG-180 turbojets located immediately behind the piston engines. These modern configurations were starting to have the design appearances that culminated in the P5Y and R3Y. The flying boats in this study were sized to 85,000 pounds. (Convair via SDASM)

The second configuration studied featured twin floats, was also sized to 85,000 pounds gross weight, and also increased the wing loading to 60 pounds per square foot from the 45 pounds per square foot, ground ruled for the other configurations. This resulted in a somewhat smaller aircraft, an increased top speed (4 percent), and a somewhat reduced range. The alternate configuration that maintained the 45-pound wing loading had generally inferior performance. (Convair via SDASM)

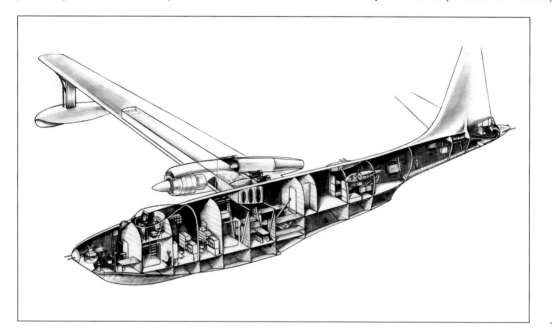

The gun turret positions can be seen in this cutaway drawing in the nose, tail, and two aft-fuselage positions. Radar operator and bombardier positions were in a bow compartment shared with the forward fire control, navigator, and bow gunner positions. The TG-180 jet engine installation was placed just behind the piston engine and on the top of the wing. (Convair via SDASM)

As World War II concluded and the technology breakthroughs in the area of gas turbine propulsion began to bear fruit, design studies were conducted to explore the benefits for military missions. One of the first studies concerned the use of compound propulsion; that is, the use of the new turbojet engines in combination with piston engines. This particular study, completed in December 1944, investigated three different airframe approaches to a patrol reconnaissance mission using two Pratt & Whitney R-4360 piston engines and two GE TG-180 turbojets. The first configuration was a conventional seaplane design, the second configuration utilized twin floats, and the third included a fairly unique single-float design. All three were sized to 85,000-pound gross weight and a wing loading of 45 pounds per square foot.

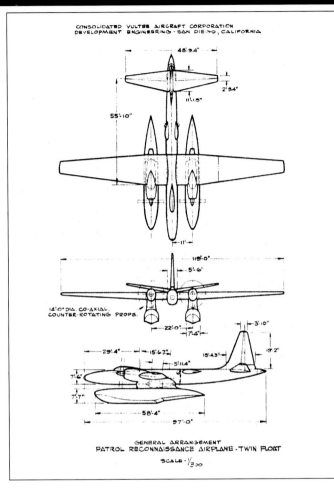

CONSOLIDATED VULTEE AIRCRAFT CORPORATION
DEVELOPMENT ENGINEERING · SAN DIEGO, CALIFORNIA

GENERAL ARRANGEMENT
PATROL RECONNAISSANCE AIRPLANE - TWIN FLOAT

SCALE · 1/300

The twin-float configuration had a wingspan of 119 feet, a length of 97 feet, and a wing area of 1,415 square feet. Propulsion was the same on all three versions, two R-4360s and two TG-180 turbojets. It can be noted that the turbojets were installed under the piston-engine nacelles in the float strut support structure. No explanation concerning the operability of the jet in the potential high-spray environment was offered. (Convair via SDASM)

Crew accommodations and equipment appear to be generally similar for all versions, although some rearrangement can be seen. The fuselage was much slimmer in this version; most of the fuel was carried in the floats. The pilot and copilot were in a tandem cockpit rather than side-by-side as in the conventional version. The bombs in this configuration were relocated from the wing to the fuselage because wing-carried stores were complicated by the float arrangement. (Convair via SDASM)

Seaplane Programs

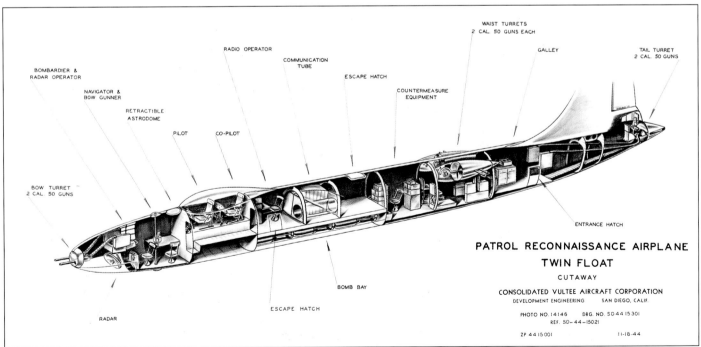

PATROL RECONNAISSANCE AIRPLANE
TWIN FLOAT
CUTAWAY
CONSOLIDATED VULTEE AIRCRAFT CORPORATION
DEVELOPMENT ENGINEERING   SAN DIEGO, CALIF.
PHOTO NO. 14146   DRG. NO. SD 44 15301
REF. 50-44-15021
ZP 44 15001   11-18-44

M-23 BOMBSIGHT — PILOT · BOW GUNNER — NAVIGATOR — STARB'D. GUNNER
RADIO OPERATOR

COUNTERMEASURE-RADAR OPERATOR — PORT GUNNER — TAIL GUN AMMO. CANS
BOMBARDIER · RADAR OPERATOR
CO-PILOT
RADAR SCANNING UNIT — DEPTH BOMBS - 8 · 650" — SIDE GUN AMMO. CANS
ASTRO-DOME — LIFE RAFTS

BOW GUNS — GALLEY — ENTRANCE HATCH — TAIL GUN INST.
ANCHOR DOOR — LAVATORY
BOW GUN AMMO. CANS — CREWS' QUARTERS — TAIL GUNNER

The installation of the jet engines can be seen in this view below the upper tail boom fuselage and in the wing pylon support structure. Depth bombs were again carried in the wing, as were the fuel tanks. (Convair via SDASM)

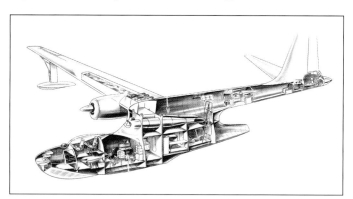

Defensive armament common to all three concepts again consisted of two turrets on each side of the forward fuselage, two waist turrets on the tail boom, and a tail turret, all with twin .50-cal. guns. The crew and the equipment compartments had somewhat different locations in the interior of the float compared to the other versions. (Convair via SDASM)

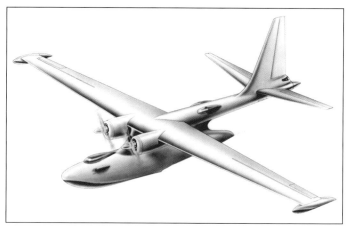

The third configuration studied was a unique single-float design with the crew and the equipment accommodation incorporated into the float itself. A small fuselage boom supported the tail. Jet engines in this arrangement were located in the wing pylon support structure, and the airplane's performance was slightly less than that with the conventional hull. (Convair via SDASM)

An alternate version of the twin-float configuration was also included that had a wing loading of 60 pounds per square foot, resulting in a slightly smaller airplane.

The performances of all of the 45-pound wing-loaded airplanes were within 5 percent, with the twin-float airplane having less performance. Since the study of that version was completed first, it probably resulted in looking at a higher wing loading design that

Seaplane Programs

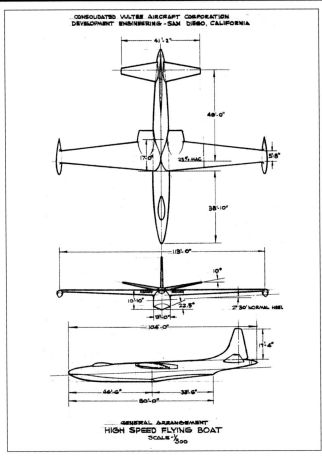

CONSOLIDATED VULTEE AIRCRAFT CORPORATION
DEVELOPMENT ENGINEERING - SAN DIEGO, CALIFORNIA

GENERAL ARRANGEMENT
HIGH SPEED FLYING BOAT
SCALE · 1/300

This jet configuration was 104 feet long, had a wingspan of 113 feet, and had a gross weight of 90,000 pounds. It was in the same general size category as the compound-propulsion versions studied a few months earlier. This study was essentially a seaplane version of what was to become the B-46 Medium Bomber that was successfully proposed in November 1944. The principal differences were the location of the jet engines and the use of the Navy Westinghouse jets. (Convair via SDASM)

might be more competitive. That version resulted in a higher maximum speed, about 4 percent, but a range decrease of 5 percent.

By May 1945, and not long before World War II ended, Convair's first study of an all-jet seaplane had been completed. This High Speed Flying Boat had a gross weight of 90,000 pounds and was to be powered by six Westinghouse 24C turbojets having a maximum thrust of 3,000 pounds each. In general this study was inspired by the work on the XB-46 with the obvious design changes necessitated by water basing. No other information concerning this study survived. Some of the design features and the "look" of this jet seaplane and of the compound propulsion flying boats discussed above were starting to have the appearance of the later P5Y and R3Y flying boats of the early 1950s.

This was the first turbojet-powered flying boat that Consolidated studied and was to be powered by six Westinghouse 24C turbojets having a maximum thrust of 3,000 pounds and a normal thrust of 2,430 pounds. The six turbojets were installed in the wing at the root with their intakes in the wing leading edge. Wingtip floats appear to have been fixed. (Convair via SDASM)

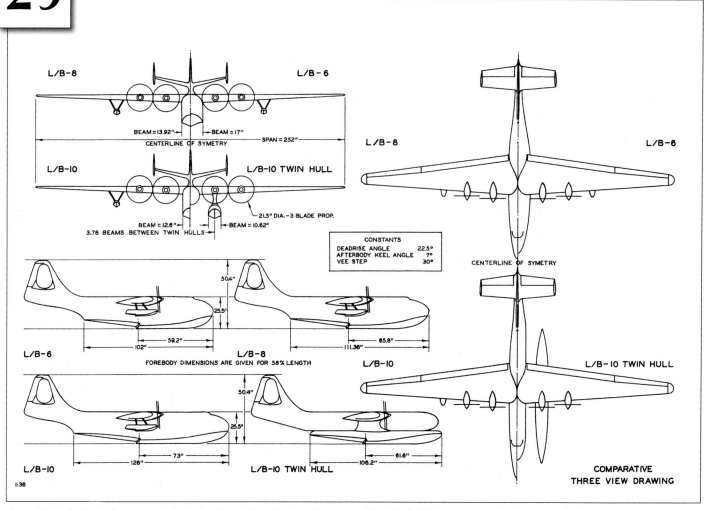

Seaplane Programs

Consolidated adopted a research tact starting during the development of the Model 31 seaplane of using dynamically similar models of the specific configuration being studied to determine the hydrodynamic characteristics of the full-size aircraft. The Navy sponsored much of this work and, in this case, variations in the hull L/b were being investigated. Variations of the L/b from 6 to 10 can be seen in these drawings of a generic four-engine seaplane. (Convair via SDASM)

In the pre-World War II era, Consolidated Aircraft had started applying innovative research techniques to the seaplane development process. The company was using dynamically similar scale models as early as 1938 under the direction of Ernest Stout. Stout believed that these techniques allowed Convair to depart from the then-prevalent "boat with wings" approach. He spearheaded the effort to develop and utilize the concepts of the aeronautical sciences, as opposed to the tenets and philosophy of naval architecture; that is, a highly refined displacement-type hull approach for seaplane development.

The use of these free-body dynamically similar models continued to be refined in the work done after the war. These techniques included the radio-controlled free-flight models, sheltered water model towing, as well as catapult model launchings, and an indoor tow basin. Based on the premise that jet power would permit seaplanes to compete performance-wise with landplanes, the Convair Hydrodynamic Research Group, and the Hydrodynamic Labs, was formed in 1943. Under the sponsorship of the Bureau of Aeronautics, a contract (BuAer NOa[s]-2754) was awarded to Convair, aimed at developing practical and competitively performing

T-4463   4-16-45
CONSOLIDATED VULTEE
1-10 SCALE RADIO FREE
FLIGHT MODEL-AERODYNAMIC
TEST SET-UP-GENERAL VIEW

This model of a generic four-engine seaplane built in 1945 had a wingspan of 21 feet and was intended for both captive and free-flight testing. This model is shown in a captive test installation mounted on a company "woody" station wagon. No information is available as to the test procedures, but it is presumed that test runs were made on airport taxiways at various speeds with pertinent aerodynamic parameters being recorded. (Convair via SDASM)

water-based airplanes through a comprehensive, basic experimental research program. Model testing had already successfully demonstrated the ability to determine the hydrodynamic characteristics of the prewar Model 31 and, later, a generic four-engine flying boat, and was utilized in the initial design activities for the P5Y and R3Y development programs in the early 1950s.

Examples of this continuing research on seaplane configurations, suggested by the generalized seaplane study, are exemplified in configurations examining implications of varying hull L/b ratios. In this case, design configurations with L/b ratios of 6, 8, and 10 were analyzed for the characteristics and effects on a large four-engine flying boat. As in many specific research studies, large models were built and tested as either free-flight models and/or as towed models for developing hydrodynamic characteristics. This particular seaplane model was a powered free-flight model. It also underwent aerodynamic testing mounted on an automobile that presumably made test runs on the airport taxiways.

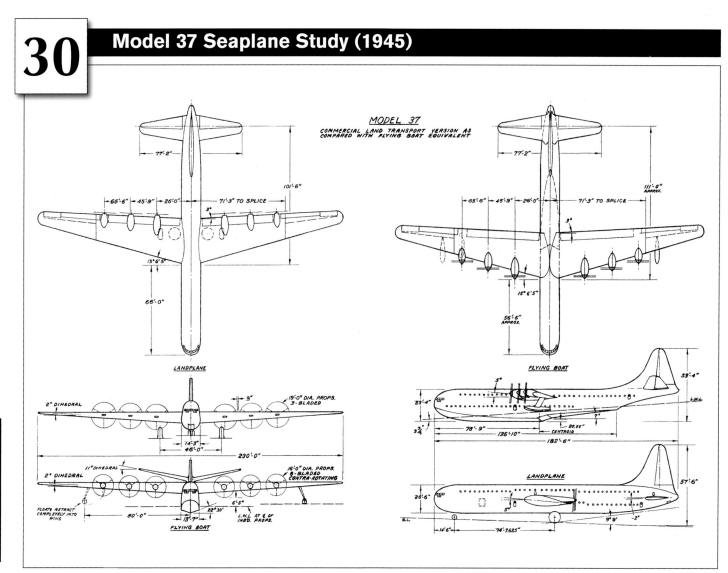

Pan American World Airways had ordered 15 commercial passenger versions of the Model 37 (the B-36 transport conversion) early in 1945. Pan American Airways requested an informational comparison of a seaplane version of this aircraft with the land-based design. The seaplane retains the same general size but substitutes a seaplane fuselage/hull. Its wing is identical to the Model 37 but engines are reversed to a tractor arrangement so as to maintain adequate propeller clearance from the spray pattern. (Convair via SDASM)

In 1942, after studies had been made of cargo and troop transport versions of the XB-36 (Model 36), the Army Air Force ordered a single XC-99 (Model 37). Commercial versions of passenger airplanes were also investigated and discussed with the airlines and PAA in particular. PAA went so far as to order 15 of these 204-passenger airliners in February 1945 with production to begin at war's end. The program was never initiated and its demise was probably due to economic considerations and the advent of the DC-6/7 and Constellation airliners.

PAA had also asked Convair for a comparison between the landplane Model 37 and a seaplane version.

The results of this study in August 1945 showed a flying-boat configuration that was generally similar in size and layout to the land-based Model 37 aircraft. The fuselage incorporated a seaplane hull and the engine arrangement had been reversed to a tractor arrangement to maintain the necessary water clearance for the propellers. Turbine propulsion technology was well underway at the time and turboprop engines were considered for this design.

It is not clear how interested PAA was beyond just informational data, and this concept was studied only very briefly.

Seaplane Programs

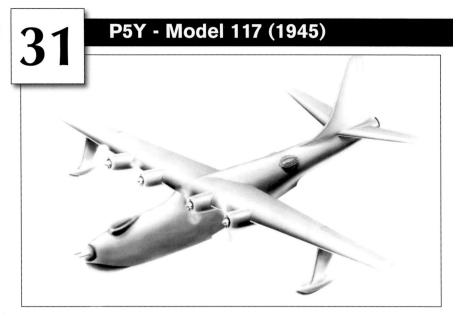

Convair's entry of December 1944 in the Navy's competition for a 105,000-pound long-range patrol bomber established the design that would carry through to the XP5Y-1 and the R3Y. Four R-2800 engines powered this plane but resulting performance discouraged the Navy from continuing this particular program. *(Convair via SDASM)*

The profile of this patrol bomber is familiar from later studies and programs. It appears that there was a crew of seven. Defensive armament included four turrets, nose, tail, and one on each side of the fuselage, each with twin .50-cal. guns. Side turrets and the tail turret were of Convair design and an Emerson ball turret was used in the nose. Note that the plane's bombs were located in the wing as with earlier designs. (Convair via SDASM)

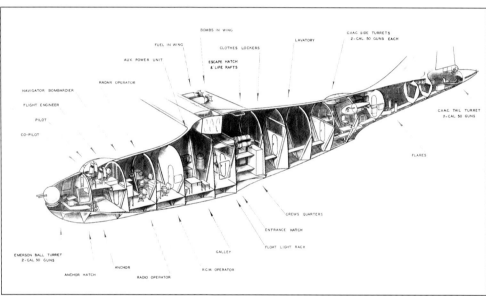

Seaplane Programs

Convair had received a Navy research contract in 1944, proposed by Ernest Stout, to study large seaplanes with enhanced performance, including hull design work aimed at improving seaplane hydrodynamic characteristics. Several design studies conducted in 1944 to 1947, including the compound-propulsion studies, showed a continuing evolution of the Convair seaplane look, eventually culminating in the P5Y/R3Y seaplanes.

In December 1944 the Navy held a competition for a Patrol Bomber seaplane powered by four Pratt & Whitney R-2800 engines having a 4,000-pound bomb capability. The mission application of this aircraft was to include anti-shipping, ASW, and search-and-rescue, in addition to the basic patrol missions. Convair responded in January 1945 with the specified piston-powered version but also submitted an alternate turboprop-powered configuration a short time later. The two configurations used nearly identical airframes and the conventionally powered version used the Pratt & Whitney R-2800-14 single-stage, single-speed 2,100-hp engine. The turbine-powered version used the Wright GTAA 5,000-hp turboprop engines with regeneration. The airframe for these seaplanes had the same 2,625-square-foot wing. The turbine-powered design, however, was significantly heavier with a gross weight of 140,000 pounds versus 105,000 pounds for the piston version. The design features included in these seaplanes are a clear predecessor to the eventual P5Y design.

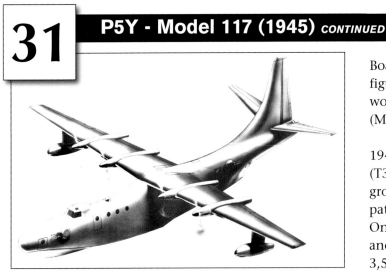

The XP5Y-1 had been proposed with the Westinghouse 25-D (T30) turboprop engines. When the Westinghouse engine was canceled and the T40 substituted, a quite-different nacelle design resulted. Defensive armament included twin 20mm gun turrets on the fore and aft fuselages, one on each side, plus a tail turret. Other than engines, the proposal airframe was the same as was built for the P5Y. (Convair via SDASM)

The Navy apparently felt that the piston-engine seaplane specified in the competition offered too little performance improvement and that particular solicitation was not pursued further. The mission remained, however, and a BuAer requirement was issued dated 27 December 1945, and a competition was initiated for a 165,000-pound four-turboprop Long Range Patrol Flying Boat generally similar to Convair's alternate study configuration previously submitted. Convair subsequently won the new award with its proposal for the XP5Y-1 (Model 117) in competition with Martin and Hughes.

The Mockup Inspection was held on 1 December 1946. The P5Y-1 design revealed a Westinghouse 25-D (T30) turboprop-powered flying boat with a normal gross weight of 123,000 pounds intended for normal patrol, search-and-rescue, and anti-submarine missions. On normal patrol it had a combat radius of 1,170 miles and a maximum range of 2,340 miles at 265 mph or 3,540 miles at 165 mph. Its maximum speed was 291 mph and the maximum endurance was 23.4 hours. The defensive armament included five remote-controlled turrets, one on each side of the fuselage in fore and aft positions and one in the tail position. Each turret carried twin 20mm guns. The bombs were carried preloaded on racks internally in the wing at the trailing edge between the fuselage and the inboard engine. These internal bomb bays had a capacity of 4,000 pounds of stores.

Gas turbine propulsion was well underway by this time and many companies had launched development programs, but the basic technology, especially the turboprop, had not yet had enough time to develop a level of maturity. This situation was to plague the XP5Y-1 and the subsequent R3Y throughout the duration of those programs. The XP5Y-1's intended engine, the Westinghouse T30 turboprop, was encountering serious development problems and ended up being canceled in February 1947. The Navy then substituted the Allison XT40-A-4 that was still in the experimental stage of development and had serious problems of its own. The powerplant was a continuing source of worry and

The proposal engines, Westinghouse 25-D (T30) geared turboprops, were to be mounted to the rear of the wing and the propellers driven by means of extension shafts that resulted in a very clean installation. Also of note was the asymmetrical engine inlet air duct located on the wing leading edge outboard of the engine installation. (Convair via SDASM)

LEFT HAND INBOARD NACELLE

DOORS IN NACELLE TOP FOR POWER PLANT INSTALLATION

DOORS IN WING FOR FUEL CELL INSTALLATION

ACCESSORY SECTION FIREWALL

TURBINE SECTION FIREWALL

STARTER

ACCESSORY COOLING AIR

GENERATOR

EXTENSION SHAFT

ANTI-ICING HOT AIR TO LEADING EDGE OF WING

OIL TANK

WESTINGHOUSE 25D PROP. TURBINE

PROP AND GEAR BOX MOUNT SUPPORTED ON MONOCOQUE STRUCTURE

FLAK PROTECTION

HYDRAULIC PUMP

OIL COOLER AIR EXIT TO SHROUD COOLING DUCT

ANTI-ICING AIR DUCT

AIR INLET DUCTS

GEAR BOX

OIL COOLER

The XP5Y-1 first flew in April 1950 about 18 months later than scheduled because of continuing development problems with the Allison XT40 turbine engine. The XP5Y-1 is shown landing in San Diego Bay during developmental flight testing. (Convair via SDASM)

Convair seriously proposed an R-3360 powered version for one of the test aircraft. This was at first approved by the Navy but then the decision was reversed.

The second situation impacting this program had to do with Navy requirements. World War II had ended and the mission and its requirements were greatly diminished. Prior to the advent of the Cold War there were no requirements in any of the projected missions for large flying boats, namely for long-range patrol, mine laying, or ASW. The engine problem and this lack of a mission combined to cause a significant program delay of about three years. Convair persistence, however, did prevail in attempts to continue this project and it eventually ended up as the R3Y seaplane transport.

The XP5Y-1 contract (NOa[s]-8347) called for one stripped test aircraft and one complete with combat equipment. The design was generally based on that developed in the December 1944 turboprop configuration but with a redesigned narrower hull with a higher L/b ratio. The fuselage had no interior bulkheads, which was a distinct advantage when the program was converted to a transport. The general tail form was adopted from the B-46, and it used fixed floats near the wingtip. The XP5Y-1 had a wingspan of 145 feet 9 inches, a wing area of 2,102 square feet, and a length of 127 feet 11 inches. The normal gross weight was 123,500 pounds and it was to be capable of carrying 8,000 pounds of bombs. The four Allison XT40-A-4 engines were to have developed 5,100 hp each that provided a cruise speed of 225 mph and a maximum speed of 388 mph. Combat radius with eight 325-pound depth charges was 2,785 nautical miles (nmi). The two XP5Y-1 aircraft were completed in December 1948 and June 1949, respectively, albeit without engines. Because of the severe engine problems, the first flight was delayed until 18 April 1950, about 18 months late.

In this period of powerplant delay Convair looked at several variants of the XP5Y-1. A long-range cargo transport version was the first logistics mission studied for this airframe and was to be powered by four Pratt & Whitney R-2800 engines, undoubtedly considered due to the T40 engine problems. This design eventually led directly to the R3Y transport that did, however, use the T40 engines.

Another design study in 1949 was a two-engine version of the XP5Y-1 using Wright R-3350 compound engines for an ASW mission. When these larger engines came on the horizon, Convair had looked at many of its four-engine airplanes with an eye to converting them to twin-engine versions with improved performance. In this study, the fuselage of the XP5Y-1 was shortened slightly and the design used a somewhat smaller wing and horizontal stabilizer. Otherwise the airframe was visually identical. The armament of Mk41 or Mk35 torpedoes was carried in the rear underside of each of the engine nacelles. Bombs and mines could also be accommodated.

Seaplane Programs

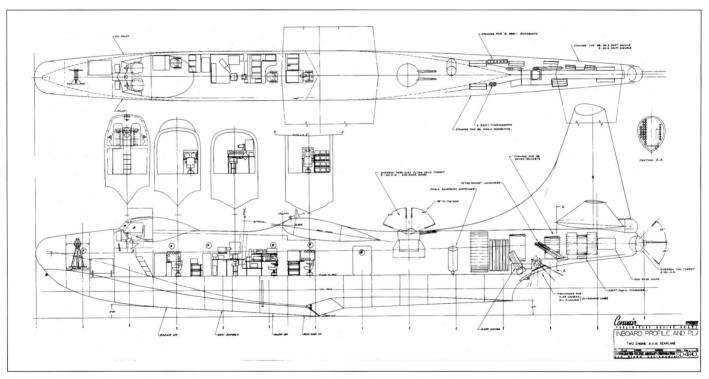

TWIN ENGINE ASW FLYING BOAT

Convair
SAN DIEGO, CALIFORNIA

During the period of the XT40 engine delay, alternate piston-engine missions were investigated. In 1949, this twin-engine P5Y was configured as an ASW seaplane using two Wright R-3350 compound engines. When these larger engines became available, this approach was used with several of Convair's four-engine aircraft to offer improved performance. The twin-engine ASW was a slightly smaller version of the XP5Y-1 but retained many of the design features and used much of the same structure. Wingspan was 116 feet versus 145 feet 9 inches for the XP5Y-1, and the fuselage was about 20 feet shorter as well. Wing area was 1,350 square feet compared with 2,102 square feet for the XP5Y-1. *(Convair via SDASM)*

The ASW gear and its operators were located on the main deck between the cockpit and the wing. Defensive armament on this ASW aircraft consisted of an upper aft-fuselage flush-deck turret (Emerson Aero X14A) and a tail turret, also Emerson. Both turrets carried twin 20mm guns. *(Convair via SDASM)*

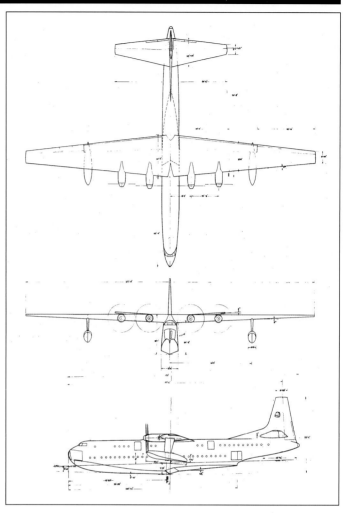

This passenger and cargo seaplane named Portage was studied in May 1947. It was a large and quite conventional configuration powered by turboprops; it is unknown if this Navy study was a response to a specific requirement or if it was a general exploratory study to determine the benefits of the new turbine propulsion. The overall design shows definite characteristics that tend to define the later R3Y, for instance the empennage and the high fuselage L/b of 10. (Convair via SDASM)

This flying boat had a large 4,200-square-foot wing with a span of 213 feet and a length of 157 feet 11 inches. It was designed to carry 50,000 pounds of payload for 5,000 miles at 300 mph, and was to be powered by four Wright T35-3 turboprop engines. The Portage had a gross weight of 315,000 pounds and a useful load of 191,900 pounds. The double-deck arrangement accommodated 200 passengers and a crew of five. (Convair via SDASM)

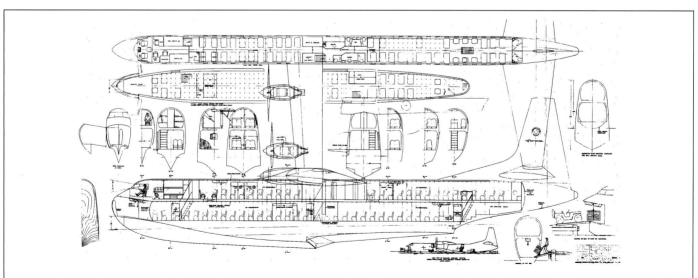

The two-deck passenger accommodations were somewhat hampered, at least by today's standards, by the high L/b of 10 that only allowed four (2+2) seats per row. Passenger loading was via a door at the upper deck in the aft fuselage. For the cargo mission there were four large fuselage doors (two on each side), in addition to the bow loading provisions for rapid cargo loading. (Convair via SDASM)

In the spring of 1947, after the P5Y was under contract but long before its first flight, Convair conducted a preliminary design study of a Long Range Personnel and Cargo Transport Flying Boat. The study report, dated 6 May 1947, documented this work conducted under Navy contract NOa[s]-7919. This seaplane, occasionally called the Portage, was to be capable of transporting 200 personnel or 50,000 lbs. of cargo 5,000 miles. The Portage was a large aircraft designed for quick conversion between cargo and passenger use and had four large fuselage doors as well as a nose door for rapid cargo loading and unloading. Four Wright Aeronautical Corporation T35-3 turboprop engines were selected as the powerplants for this seaplane. The T35, early on the technology curve, was to have provided about 7,800 hp but because of development problems never became available.

33

Assault Seaplane Transport (1948)

AIRPLANE GENERAL ARRANGEMENT

REF: BD-47-17010

In 1948 Convair studied an Assault Seaplane Transport (AST) for the Navy that was to evaluate an efficient seaplane for long-range missions to transport cargo. The overall size of the AST was about 50-percent larger than the P5Y but it retained the same overall appearance except, of course, for the transport type fuselage that was a single-deck design. It had a wingspan of 236 feet, length of 197 feet, and gross weight of 360,000 pounds. Because of a lack of documentation on this project, the performance and transport capabilities are unknown. The type of the four turboprops that powered the AST is also unknown but they drove six-blade, counter-rotating, 19-foot propellers, as compared to the 15-foot props on the P5Y. (Convair via SDASM)

A 15-percent scale model was used to investigate hydrodynamic stability and to determine spray patterns. A powerboat towed the model in both open and sheltered water. The hull, based on the P5Y design, had an L/b of 10, which Convair determined as the most efficient from past research studies. The model at this point omitted the engines and the original-design wingtip floats. (Convair via SDASM)

Another large seaplane transport study was conducted in 1948 for the Navy under contract NOa[s]-9123. This study for an Assault Seaplane Transport (AST) was addressing the concern growing out of the increasing submarine-warfare threat to the Navy's logistics and support operations. It was postulated that a large, efficient seaplane capable of transporting large bulk cargo over long distances would be a promising counter for this threat. The study yielded a large flying boat with a gross weight of 380,000 pounds, a 5,070-square-foot wing, and a span of 236 feet. The configuration generally resembled the yet-to-be available R3Y-2 but was significantly larger, by about 50 percent. It was pointed out that the P5Y fuselage/hull represented a 2/3-scale AST. A 1/15-scale dynamic free body model of the AST was constructed and tested both as a towed-sheltered model and an open-water model to determine the hydrodynamic stability and spray characteristics. It was also used to investigate the operability of the design with regard to operational support and the assault mission.

Seaplane Programs

MACH 0.90 - LONG RANGE SEAPLANE BOMBER

FOREBODY ONLY

COMBINATION AFTERBODY AND OUTBOARD FLOAT

This VA Long Range Special Attack airplane is a tailless design that used a combination of floats and vertical stabilizers at the wingtip. The jet engines were mounted on the rear of the fuselage and the induction intakes were on the top of the fuselage. This airplane performed in the high-subsonic-speed range (Mach 0.9) and was in the 100,000-pound gross weight class. (Convair via SDASM)

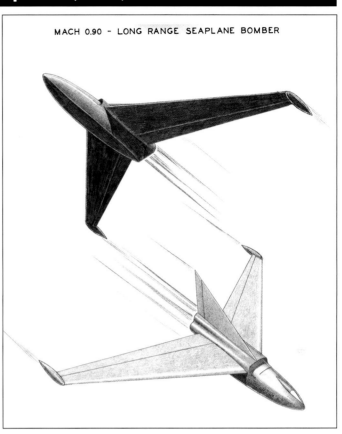

An alternate design for the Mach 0.9 long-range seaplane bomber retained the tailless swept wing but used a single vertical fin at the fuselage rather than the wingtip verticals. The designs in this section are from the proposal and can be considered only preliminary examples. (Convair via SDASM)

Seaplane Programs

In August 1948 the Navy issued Outline Specification OS-111 and OS-115 for a Class VA Long Range Special Attack Airplane, an early look at a nuclear weapon capable aircraft. This supersonic nuclear-strike weapon was intended to operate from aircraft carriers and would be in the 100,000-pound class of aircraft. Convair responded and conducted a study of a two-component system that jettisoned sections that were not essential for the return of the crew. This conceptual approach was probably inspired by the ongoing USAF GEBO (GENeralized BOmber) studies that had included the composite-aircraft two-component approach. These studies had been initiated in October 1946 at the Convair Fort Worth Division.

The application of this earlier two-component concept showed up again when Convair was informally requested by the Navy in March 1949 to suggest types of aircraft around which future additional work could be undertaken in the high-Mach-number seaplane section of the ongoing Skate research contract. The OS-111 and OS-115 requirements were used as a basis for the transonic Mach 0.85 and supersonic Mach 1.2 water-based aircraft to be investigated as part of the subject follow-on program. One of the goals of Convair's research was to base the designs on real-world requirements. This research would be aimed at an attack seaplane whose performance goals were sufficiently advanced to make possible its development prior to the appearance of comparable landplanes. Convair proposed a tailless swept-wing configuration for the M.85 aircraft and a delta-wing composite aircraft for the supersonic mission. Meanwhile, in June 1949, the Navy had

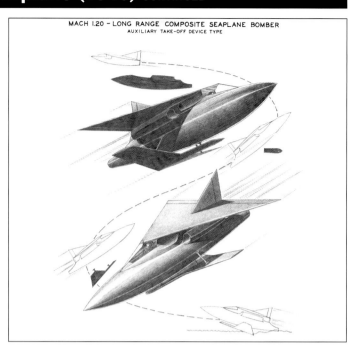

MACH I.20 - LONG RANGE COMPOSITE SEAPLANE BOMBER
INTEGRAL HULL BOTTOM DESIGN TYPE

MACH I.20 – LONG RANGE COMPOSITE SEAPLANE BOMBER
AUXILIARY TAKE-OFF DEVICE TYPE

The supersonic (Mach 1.2) seaplane bomber represented a two-component composite airplane and had an integral hull-bottom design. The water-based takeoff and initial cruise was made with both components and at some point after the takeoff those components were jettisoned and the remaining component completed the mission supersonically and then returned to its base. (Convair via SDASM)

An alternate suggestion for the Mach 1.2 two-component configuration added an auxiliary takeoff device. This device appears to be a very early concept for a form of hydro-ski. At this point of the technology there was very little research available to establish a viable design. This device was jettisoned immediately after takeoff and the airplane then proceeded to the point where the second component was also jettisoned, as in the first system. (Convair via SDASM)

requested that NACA and Convair investigate the application of hydro-skis to the Skate fighter concept. The Convair response concluded that the blended wing-hull was more efficient and preferable for these specific requirements but the possibilities inherent in the hydro-ski concept justified further study.

As it turned out, the objective of this follow-on program was to develop a practical water-based aircraft similar to the Class VA Long Range Special Attack requirements. It was decided that the supersonic attack seaplane program should investigate the blended wing-hull as well as the hydro-ski approach for the various tactical requirements. It should be borne in mind that all of the work on the Cudda and the Betta, as they were named, was conducted when the hydro-ski was viewed as having very great potential. The work took place at least two years before the SeaDart first flew and revealed the serious problems with the hydro-ski concept.

The design and experimental program was fashioned around the concepts developed in March 1949 for the follow-on study contracted under BuAer NOa[s]-9722. These studies included:

I. Transonic concepts (BuAer Spec OS–111, July 1948)
    A. Blended wing and hull type
       (Skate, already underway)
    B. Hydro-ski type (Cudda)
II. Supersonic concepts (BuAer Spec OS-115,
      August 1948)
    A. Blended hull type (Betta 2)
    B. Hydro-ski type (Betta 1)

This proposal led directly to the Cudda and Betta studies. The two-component configuration was not pursued further.

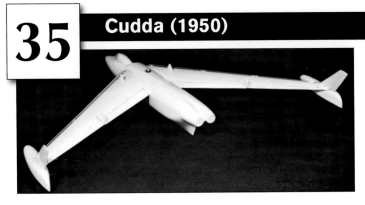

The objective of the Cudda concept, seen in this rare but low-resolution archival image, was to look at a water-based alternative to the carrier-based, heavy-attack bomber requirement of the time. Details of this Cudda configuration can be seen on this 1/40-scale hydrodynamic test model. Tailless configuration with upper fuselage engine installation is apparent. The swept wing accommodated full span leading edge slats and had wingtip floats and vertical stabilizer surfaces. (Convair via SDASM)

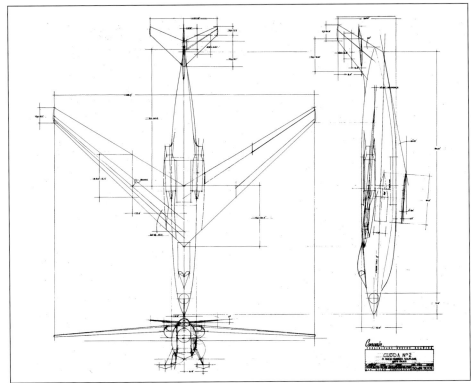

The second iteration, the Cudda 2, investigated hydro-ski applications to this concept. It retained the high-aspect-ratio wings but now utilized conventional tail surfaces, moved the jet-engine powerplants to a mid-fuselage position, and added the hydro-skis for takeoff and landing. The wingspan of this aircraft was 88 feet 8 inches and its length was 80 feet. (Convair via SDASM)

Seaplane Programs

The transonic Cudda configurations were designed around the requirements of the Navy's VA Long Range Carrier Based Subsonic Heavy Attack Bomber fulfilled by what was to be the A3D. This requirement called for the delivery of a 10,000-pound payload, 1,700 nmi at a speed of M.85 at an altitude of 40,000 feet. The name Cudda is believed to be a contraction of Barracuda, a very aggressive ocean fish. It was envisioned the Cudda would operate out of a home base but be supported from an extended base, either another airplane, a surface ship, or a submarine. The avowed purpose of the study was to apply the latest aerodynamic and propulsion technology to water-based aircraft and to incorporate radical new seaplane designs.

The initial study, conducted under contract NOa[s]-9722, centered on a tailless high-aspect-ratio swept-wing design for Cudda 1, similar to the initial proposal response in 1948. In addition, Cudda 1 featured an unconventional hull (no step) and wingtip floats with vertical fins. It was felt that the tip floats far aft on the swept wing would perform essentially as a conventional afterbody in controlling trim. The powerplants, two TJ-6s (24,000-pound thrust) turbojets were mounted on top of the wing. Extensive towline testing of a 1/40-scale model was undertaken using Convair's motor launch in San Diego Bay adjacent to its main plant.

The next design concept investigated for Cudda was a hydro-ski configuration with a conventional tail.

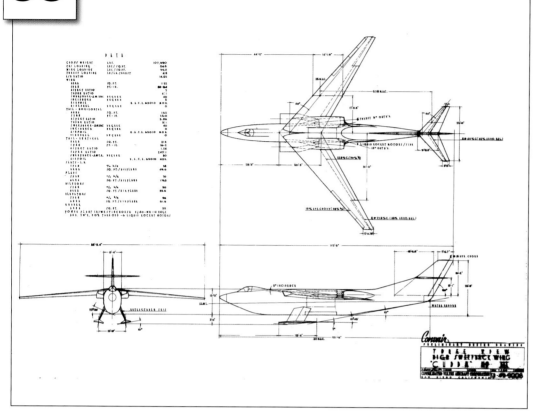

The Cudda 3 was a refined and slightly enlarged Cudda 2. The 1,122-square-foot wing had a span of 88 feet 4 inches and the aircraft was lengthened to 112 feet 3 inches. Engine installation was basically the same as the earlier version but the engines were now Westinghouse XJ40-14s. There were also provisions for six liquid-propellant rocket engines for jet assisted takeoff (JATO) assist. (Convair via SDASM)

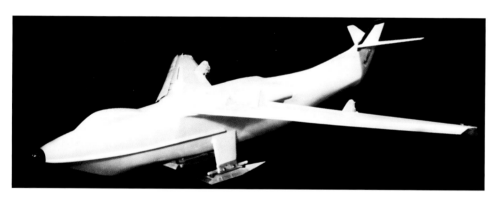

The 1/40-scale hydrodynamic tow model illustrates the hydro-ski installation and its attachments to facilitate testing a variety of shapes. It is believed that this particular design of hydro-ski was unsuccessful. Note the open leading-edge slat on the right-hand wing. (Convair via SDASM)

The aerodynamics were further refined to a cleaner, more streamlined hull, and the powerplants were moved to a position under the wing next to the hull. The Cudda 2 had a 1,120-square-foot, 40-degree swept wing, a gross weight of about 115,000 pounds, and a span of 86 feet 8 inches. Considerable testing was then undertaken, both wind tunnel and with 1/40-scale hydrodynamic models. Engines for this version were two Westinghouse XJ40-WE-14 plus six liquid rocket motors for takeoff.

Cudda 3 featured a further-revised hull with redistributed hull proportions to correct a nose-down trim while moving at low speeds in the water encountered by the Cudda 2 configuration. It retained a majority of the other features including the twin-ski approach. This configuration was to have a gross weight of 107,990 pounds.

Powerplants for this version were again two Westinghouse XJ40-WE-14s plus six liquid rocket Jet Assisted Take Off (JATO) motors. Although both Cudda 2 and 3 employed twin skis, some testing of a single-ski version was also conducted.

Seaplane Programs

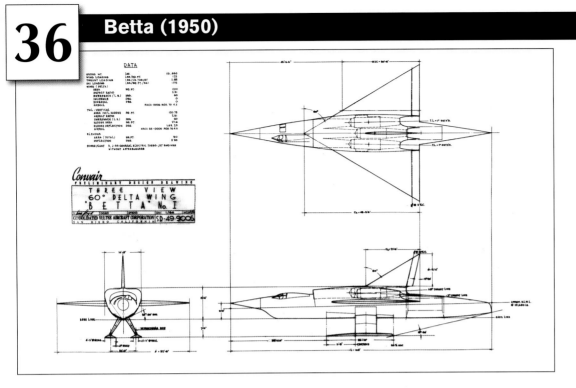

Seaplane Programs

The Betta concept was investigated using the same requirements as for the Cudda but in this case the airplane was to be supersonic. The Betta 1 design had a delta wing of 1,210 square feet, span of 52 feet 6 inches, and length of 105 feet. It utilized hydro-skis for takeoff and landing, had a gross weight of 151,650 pounds, and was powered by three J53 turbojet engines without afterburners. (Convair via SDASM)

Initial Betta design, seen in this rare but low-resolution archival image, located the engines and their intakes on top of the fuselage. The vertical stabilizer was also located on top of the fuselage; the engines, at the midpoint of the delta wing. (A long-hull afterbody gave the appearance that the vertical stabilizer was at the midpoint of the aircraft.) The model shown is undergoing low-speed wind tunnel tests at Convair. (Convair via SDASM)

The Betta studies, conducted under the same contract as the Cudda, and also an outgrowth of the 1948 proposal, were undertaken to investigate problems attendant to the creation of an advanced supersonic attack seaplane. Again the requirements called for a 100,000-pound airplane that would carry a 10,000-pound payload and have a combat radius of 1,700 nmi. Although the composite two-component aircraft had been initially proposed, predesign studies of the mission concluded the requirements could be met by relaxing the gross weights limitation of 100,000 pounds and this would avoid the complexity of the two-component concept. Accordingly the initial design configuration for the Betta 1 featured a 60-degree delta wing,

Seaplane Programs

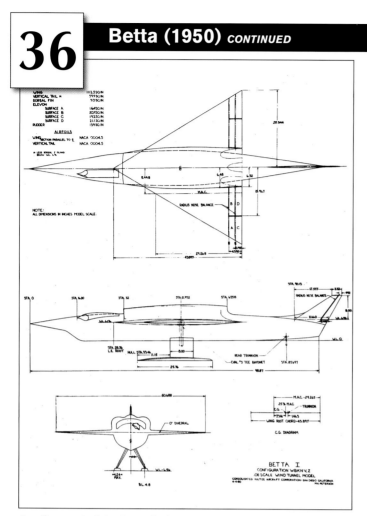

A refinement of the Betta design changed to a conventional vertical stabilizer located at the aft end of the fuselage, increased its full-scale length to about 111 feet. It retained the overwing fuselage intakes. *(Convair via SDASM)*

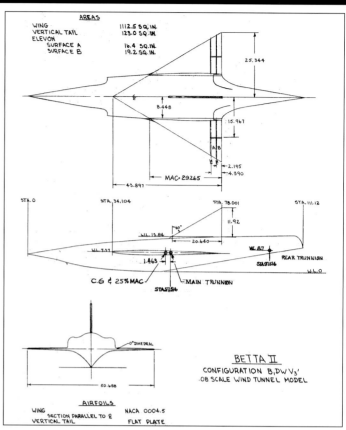

Research on the Betta 1 led to a Betta 2 configuration with side engine inlets forward of the wing leading edge and fuselage engine exits at the trailing edge of the wing. This possibly indicates a change to a twin-engine configuration although no direct information is available. Also it appears that a blended lower hull had been incorporated. *(Convair via SDASM)*

twin retractable hydro-skis, and three GE J53 jet engines without afterburners (mounted high over the fuselage). It had over-the-fuselage inlets, and an extended fuselage/hull with a relatively small vertical fin. The Betta also had an offset cockpit similar to some of the Skate configurations.

The J53 engine never completed development, but its projected rating was 18,000 pounds thrust for a total of 54,000 pounds. Wind tunnel tests were also conducted on configuration variations of the Betta 1 that included a larger delta vertical fin located at the same position as the wing, and a shortened fuselage. There is no information to indicate which configuration was the preferred one. Extensive wind tunnel and hydrodynamic testing was conducted around 1950 on both

versions. The water testing again included bridle towing with 1/40-scale models.

The Betta 2 configuration was generally similar to the Betta 1 but featured the blended wing-hull previously developed and preferred during the Skate studies, in lieu of hydro-skis. In the model exploratory phase, the wing-located vertical delta fin was featured, although several configurations were investigated in the wind tunnel, including the smaller vertical fin and extended rear fuselage, the same as the Betta 1. The Betta 2 also featured wing root shoulder inlets and wing trailing-edge engine exhausts. Wind tunnel testing of both Betta 1 and 2 was conducted in the first half of 1950 with many variations of tail and fuselage design as well as canards.

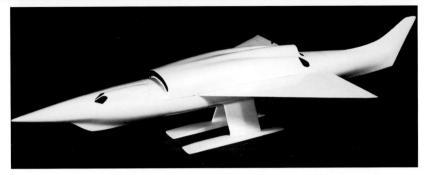

This model of the revised Betta 1 design shows the hydro-ski configuration, engine-intake inlet and exhaust, and new vertical-tail configuration. Note the pilot's offset canopy in this configuration, which provided for a second crewmember to his right completely within the fuselage. This crewmember was presumably the navigator/bombardier. *(Convair via SDASM)*

Employing both a delta wing and a hydro-ski, the Convair XF2Y-1 SeaDart was one of the most futuristic aircraft to ever take flight. Still a modern looking design today, the SeaDart actually first flew in 1953. Although not put into series production, many lessons were learned with this aircraft that assisted in the successful long-term production runs for Convair's F-102 Delta Dagger and F-106 Delta Dart supersonic interceptors several years later. *(National Archives via Dennis R. Jenkins)*

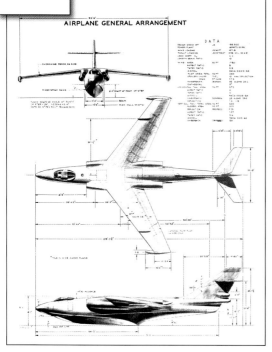

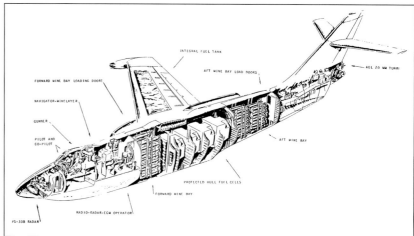

Fuselage interior of the initial design seen in this rare low-resolution archival image shows the flight deck that accommodated a crew of five in the forward fuselage. Just behind the flight deck was the forward mine bay, with the second of the two in the aft fuselage. The tail accommodated a twin 20mm tail turret. Fuel was carried in fuselage-protected cells in the mid-fuselage area between the mine bays and in integral tanks in the wing. (Convair via SDASM)

In early 1952, the Navy initiated a competition for a jet-powered seaplane and Convair, along with Martin, submitted proposals. Convair's proposal outlined a four-engine jet flying boat of generally conventional appearance. The selected J67 engines were located at the wing fuselage shoulder at the aft portion of the wing. They were staggered, one ahead of the other, with the exhausts to the side of the fuselage. The wingspan was 93 feet 6 inches and the length was 129 feet 8 inches. Gross weight was 188,500 pounds. (Convair via SDASM)

In early 1952, the Navy initiated a program for a jet-powered seaplane, the Long Range Aerial Mine Layer. The preliminary outline of the requirements guiding this competition (BuAer Outline Spec OS-125, dated 14 May 1951, as amended in June 1952) were first made available in April 1951. Both Convair and Martin initiated proposal work at that time. The requirements specified an airplane with a desired gross weight of 160,000 pounds, a speed of 600 knots at sea level and the capablitiy of carrying 30,000 pounds of mines for a combat radius of 900 nmi. The initial proposal work by Convair was conducted under the title of High Performance Seaplane. Later, in the post proposal engineering study follow-on, it was known as the Class VP High Performance Flying Boat (Aerial Mine Layer).

Both Convair and Martin submitted the proposals for an aircraft to meet this requirement in January 1952. After evaluation, BuAer decided certain aspects of both proposals required further development and identical follow-on contracts for $200,000 were issued to both companies, Convair's being NOa[s]-52-1137. These additional engineering studies were undertaken during a six-month period and culminated in extensive final proposals in August 1952.

Convair's proposal effort initially considered both hydro-skis and the blended wing-hull, but several conflicts coming out of the specifications and requirements resulted in the decision to favor hydrodynamic performance over aerodynamic form. The resulting hull configuration clearly evolved from the earlier Cudda studies. A faired, pointed-step hull similar to that commonly known as a planing tail was then adopted as the best compromise for the hull configuration. After all of the highly innovative concepts that Convair had studied, a fairly conventional configuration was selected for the proposal.

The proposal, submitted in January 1952, featured a seaplane designed to be capable of rough-water operation in waves 6 to 8 feet high. It had moderately swept wings (35 degrees), an area of 1,750 square feet, and a span of

Seaplane Programs

After the proposals were submitted, the Navy funded an additional six months of study work to both Convair and Martin for additional details and to incorporate some of the Navy's requested changes. Convair's resubmittal was visually quite similar although some changes were made, such as a slightly larger wing, engine and air intake installation changes, and mine bay revisions. (Convair via SDASM)

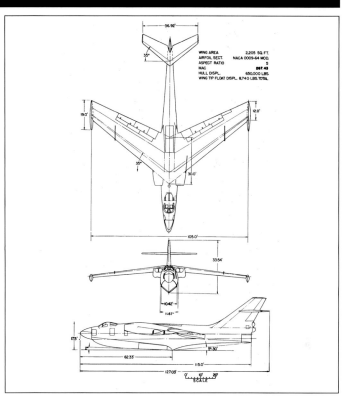

Below: In the revised configuration, now known as the Class VP High Performance Flying Boat (Aerial Mine Layer), the YJ67 (TJ32C1) engines were moved forward to a side-by-side location but were still staggered with the outboard engines forward. In addition one of the two original mine bays was eliminated, although the mine payload remained at 20,000 to 30,000 pounds. Defensive armament consisted of a radar-controlled twin 20mm tail turret. (Convair via SDASM)

The revised airplane in the final submittal was slightly larger and had a wing of 2,205 square feet and a span of 105 feet 6 inches, as compared to the initial proposal of 1,750 square feet and a span of 93 feet 6 inches in. Total length of the airplane was 127 feet. Gross weight; however, was now reduced by 14,750 pounds to 173,750 pounds. (Convair via SDASM)

Seaplane Programs

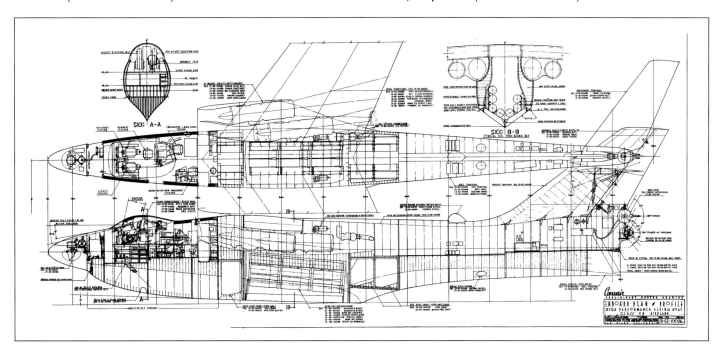

Seaplane Programs

A desk model illustrates the 35-degree swept wing and the more forward engine air intakes. The near T-tail was configured to keep the empennage out of the jet exhaust and water spray. A remote tail turret was located in the aft fuselage under the empennage. The wingtip floats also accommodated certain electronic antennas. *(Convair via SDASM)*

This view of the model shows the mine doors that open directly from the hull with the mines ready for deployment. The mine bay doors were located in the relatively low hydrodynamic impact area of the hull. The mine bay normally accommodated 20,000 pounds of mines with an alternate maximum loading of 30,000 pounds. Provisions were also included for carrying a "special weapon." *(Convair via SDASM)*

Winner of the Navy's coveted High Performance Flying Boat design competition in 1952 was arch-rival Martin Aircraft Company's XP6M-1 Sea Master. Looking eerily similar to Convair's four-engine T-tail proposal, the Sea Master held high hopes for the Navy's capabilities in long-range bombing and aerial mine-laying. However, two factors conspired to bring about the P6M's demise: Performance problems that led to the loss of two aircraft, and the Air Force's victory in ensuring that the Navy would not eclipse that service's role in strategic bombing. (National Archives via Dennis R. Jenkins)

93.5 feet. The fuselage length was 129 feet 6 inches. Four Wright YJ67 (TJ32A5) Olympus engines mounted at the wing-hull juncture powered the aircraft. This engine was a version of the British Olympus, which incorporated afterburners, and was rated at 15,400-pound maximum thrust. The engines were mounted staggered, one behind the other, under the wing next to the hull, with over-under inlets. The control system was all power operated and a conventional rudder provided directional control. Wing spoilers provided lateral control and an all-moveable hor-

izontal tail provided longitudinal control. Armament included up to 32,000 pounds of mines and a twin 20mm remote-controlled tail turret gun. The mines were stored in a forward and an aft bomb bay. This version had a maximum gross weight of 188,500 pounds and an empty weight of 96,880 pounds. It carried a crew of five.

After the follow-on engineering extension studies were concluded, the final proposal was submitted in August 1952. The revisions to the Convair configurations during these studies included a slightly reduced gross weight. The engine installation had been somewhat revised to an almost pod installation but was still located at the wing root. The location of the engines still had a staggered configuration but not as much as earlier. These engines were specified as Wright YJ67 (TJ32C1) rated at 13,500 pounds of thrust. No afterburners were incorporated as were in the January 1952 version. The two payload bays were changed to a single central bay located just forward of the main step and the design payload was 20,000 pounds of mines. A maximum capability of 32,000 pounds could be accommodated. This aircraft had a combat radius of 1,070 nmi and a high speed of 586 knots at sea level.

In the end, Convair lost the competition to Martin, which received the award in late 1952. The first XP6M-1 Sea Master was rolled out in July 1955, but Martin went on to build only a total of 16 P6M seaplanes. Development and cost problems precluded operational use of the aircraft and the program was canceled in 1959. The Sea Master was expectedly similar to the Convair configuration having a gross weight of 160,000 pounds compared to Convair's 174,000 pounds. It had a span of 102 feet 7 inches versus Convair's 105 feet. The XP6M utilized four J71 engines in the initial versions. More powerful Pratt & Whitney J75 engines were adopted for the P6M-2, whose gross weight increased to 195,000 pounds for those follow-on production aircraft.

38 Twin-Hull Seaplane (1952)

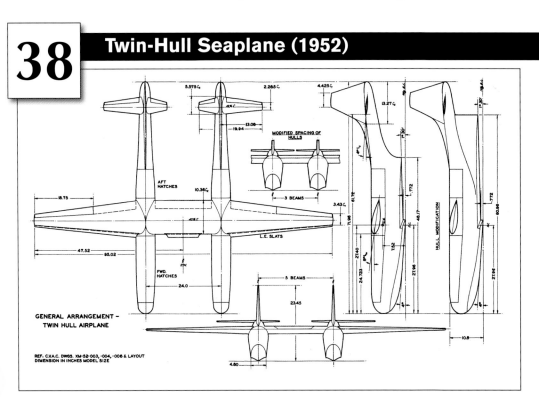

GENERAL ARRANGEMENT – TWIN HULL AIRPLANE

REF: C.V.A.C. DWGS. XM-52-003, -004, -006 & LAYOUT DIMENSION IN INCHES MODEL SIZE

Convair undertook twin-hull flying-boat research in 1952 to determine the relative merits of twin-hull and single-hull water-based aircraft. A twin-hull model was used to establish the hydrodynamic characteristics. The twin-hull version was sized for comparison with a single-hull aircraft, the R3Y. The model here investigated twin-hull spacing and a baseline for a modified hull shape. (Convair via SDASM)

Convair conducted a study under Navy research study contract (NOa[s]-12143 Amend 4) in 1952 to look at the relative merits of twin- versus single-hull water-based aircraft. As seaplanes became larger, providing transverse stability became more difficult as the size, weight, and loads of the required wing floats or sponsons increased. That then had an adverse effect on performance. The possibility was considered of eliminating them in the twin-hull configuration wherein the two hulls provided the necessary stability and the necessary propeller clearance.

The study encompassed analytical as well as experimental approaches and a dynamically similar 1/20-scale model was constructed to determine stability, spray characteristics, and landing stability and behavior. The results of twin-hull research were then compared to a single-hull design, the R3Y-1. The full-scale twin-hull airplane was sized for a gross weight of 158,000 pounds.

The resulting model shown here on the water was used in several ways including towed tests to establish spray characteristics and hydrodynamic stability. It was also used for catapult tests for landing and, again, stability behavior. (Convair via SDASM)

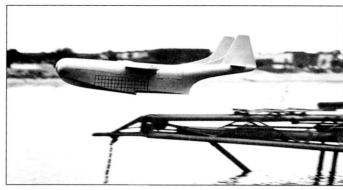

The model with the baseline hull was being launched by means of a catapult and it glided to the surface for a water landing as its aerodynamic behavior was recorded photographically. This technique was used on many of the dynamically similar model-research programs starting with the Model 31 research in 1938. (Convair via SDASM)

39   R3Y Tradewind - Model 3 (1950)

As World War II ended, the priority for Navy combat missions diminished significantly. The role of the XP5Y-1 was changed several times but because of the engine problems the program appeared to be headed for cancellation. Convair was, however, able to resurrect the aircraft as a military transport, the R3Y-1. Changes were few, centering mainly on a lengthened fuselage opened-up to better accommodate cargo and passengers, strengthened wing floats, plus a somewhat revised wing, nacelles, and empennage. (Convair via SDASM)

The mission for the XP5Y-1, a program that had serious delays because of engine problems, was changed in mid 1949 to ASW and then subsequently to mine laying. The Navy then announced in April 1950 that both combat missions were being abandoned and the program was being converted to a transport aircraft, the R3Y-1 (Model 3), and was subsequently named the Tradewind.

Design differences from the XP5Y-1 included an improved wing and strengthened wing float, an improved nacelle design that was moved more forward on the wing, and a taller empennage. The lengthened (14 feet 7 inches) fuselage featured a 10-foot cargo hatch just aft of the wing and the fuselage was fully pressurized. The R3Y-1 would accommodate 80 to 90 troops or 72 litters in the hospital configuration.

Seaplane Programs

The very limited R3Y program was expanded further with the addition of an Assault Transport version, the R3Y-2, shortly after the P5Y program was converted to a transport mission. This assault seaplane version incorporated a shortened bow loader nose that was intended to facilitate the landing of troops and equipment at unimproved beaches. In the end six of these R3Y-2s were built along with five of the R3Y-1s. (Convair via SDASM)

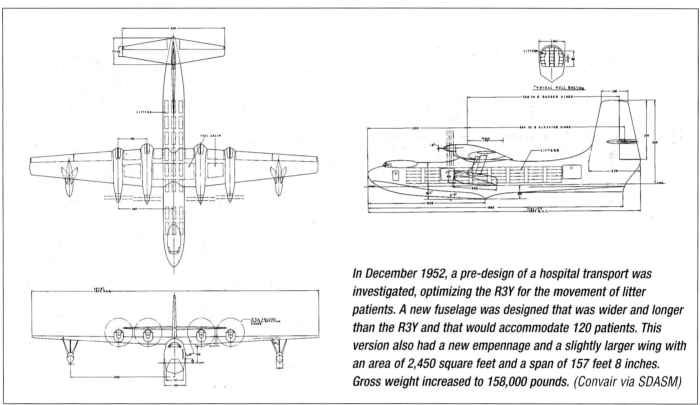

In December 1952, a pre-design of a hospital transport was investigated, optimizing the R3Y for the movement of litter patients. A new fuselage was designed that was wider and longer than the R3Y and that would accommodate 120 patients. This version also had a new empennage and a slightly larger wing with an area of 2,450 square feet and a span of 157 feet 8 inches. Gross weight increased to 158,000 pounds. (Convair via SDASM)

On 16 August 1950, six R3Y-1s were ordered and on 10 February 1950, five additional R3Y-2 aircraft were ordered. At the same time one of the R3Y-1s was converted to an R3Y-2. The R3Y-2 was configured as an Assault Transport and had a bow loading nose and a somewhat shorter fuselage. The R3Y-1 first flew using the newer Allison T40-A-10 engines on 25 February 1954, a year behind schedule, followed by the R3Y-2's first flight on 22 October 1954. Three of the R3Y-2s and one R3Y-1 were converted to probe-and-drogue tankers but were used mainly for concept testing purposes. The engine continued to be a problem although the Tradewind was used on a limited basis as a logistic

transport between Alameda NAS, California, and Hawaii. After a second significant accident attributed to engine failure, the R3Ys were grounded on 24 January 1958 and subsequently retired.

In late 1952, another variant was studied an air-evacuation hospital transport. This was basically an R3Y with a new fuselage optimized for carrying litters. The fuselage was lengthened by about 25 feet and widened to a maximum of 14 feet to better accommodate the litter configuration. This design had an entirely new vertical stabilizer and the wing had an increased span of about 12 feet. It had the same Allison T40-A-10 power-plants as the R3Y. The gross weight was slightly more, at

Seaplane Programs

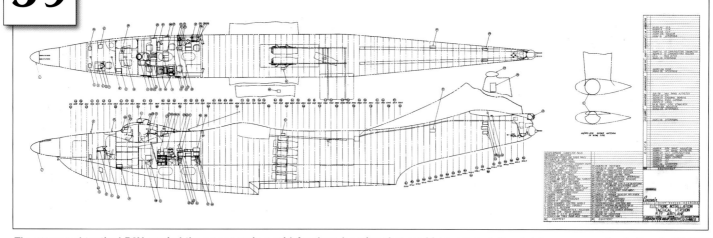

The proposed tactical R3Y carried the weapons in a mid-fuselage location that could accommodate a total weapon payload of up to 20,000 pounds. Potential stores included 2,000-pound bombs, Mk 25 mines, Mk 41 torpedo, and a special weapon, either a single 66 x 66 x 200-inch store or two 57 x 57 x 165-inch stores. Defensive armament in this version included a Westinghouse Aero X21A tail turret and an Aero X18 nose turret, both equipped with twin 20mm guns. An APS-33B radar unit was carried on the wingtips. *(Convair via SDASM)*

As late as 1958, and in the face of continuing engine problems, Convair was looking to re-engine the R3Ys with a more reliable powerplant. This approach proposed replacing the T40s with Rolls-Royce Tyne turboprops. It is not known what model of the Tyne was being considered but the rating of that engine was in the 4,500 to 5,000-hp range—quite comparable to the T40. *(Convair via SDASM)*

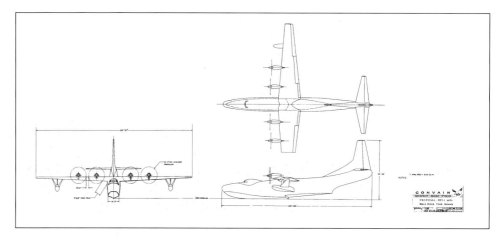

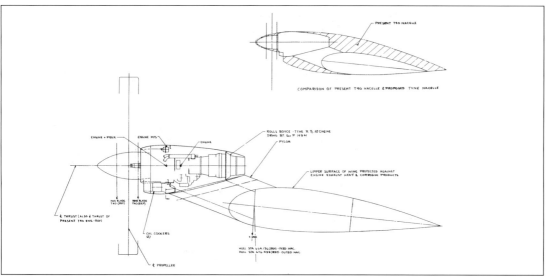

The Tyne engine installation shows a pylon mounting of the engine above the wing, placing the propellers in the identical position as they were with the T40. This would tend to minimize the overall impact this engine change had on the aircraft's performance characteristics resulting from this modification. *(Convair via SDASM)*

Seaplane Programs

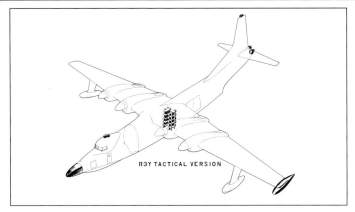

A study was conducted in June 1951 entitled "Tactical Version R3Y Airplane" that addressed an updated R3Y, including, once again, an armed militarized version. In the armed configuration, stores proposed included the conventional weapons for ASW missions and also provided for nuclear weapon delivery. Defensive armament included both a nose and tail turret. (Convair via SDASM)

Showing the original design's fixed-nose configuration, the R3Y-1 Tradewind gives an impressive sense of size and scale as it sits mounted atop special beaching gear required for ground handling on land. This apparatus was strictly for water-to-ramp and ramp-to-water transitions, and was not meant to give the giant aircraft amphibious capability. (National Archives via Dennis R. Jenkins)

Later-version R3Y-2 Tradewind poses at Convair's loading dock on San Diego Bay and, with its clamshell bow loading door in the fully raised position, offloads mock troops and a Jeep to graphically demonstrate its Combat Assault Transport capability by delivering combat assets to any port worldwide. (National Archives via Dennis R. Jenkins)

158,000 pounds versus 145,000 pounds for the R3Y.

As the Korean War was underway and the Cold War was emerging, a study was conducted in April 1951 of a re-militarized R3Y titled "Tactical Version R3Y." The uses envisioned included an attack mission with special stores, mine layer, and photo reconnaissance, in addition to the R3Y's original cargo (and cargo drop), passenger, and air-evacuation transport roles. The main visual change was a new nose-section gun turret and a unique wingtip gun turret installation. The revised assault transport version also employed a new clamshell bow loading door.

It has been reported that a study was conducted on using the R3Y as a test vehicle in some capacity on the nuclear seaplane work Convair was conducting. Application of the R3Y may have been analogous to the use of the B-36 for early reactor and shield testing, but that aspect of the program apparently did not continue beyond very preliminary studies.

A brief study was also undertaken in 1958 to re-engine the R3Y with the Rolls-Royce Tyne turboprop, presumably a more mature engine, but again no further work was done.

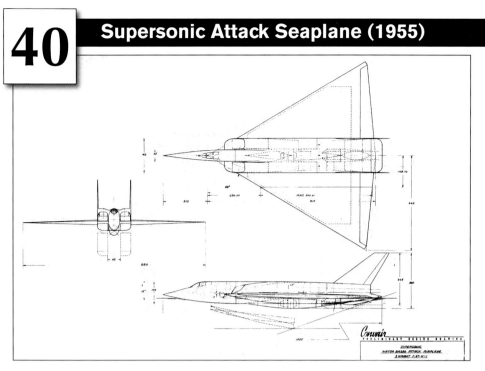

Initial configuration of the Navy study concerning a Supersonic Attack Seaplane (SAS) was developed in late 1953. This version was a delta-wing hydro-ski configuration powered by three Wright J67-W-1 engines. It had a wingspan of 74 feet and length of 88 feet. Above-wing intakes provided air induction for the three jet engines and it appears the weapon payload was carried in a semi-submerged centerline position. (Convair via SDASM)

In the final SAS report (June 1955), the first configuration was termed an F2Y-type patterned after the SeaDart as being the best solution for rough-water operations. It appears to be a refined version of the design initially studied in 1953. This was again a delta-wing hydro-ski configuration, in this case powered by two Allison 700 PD9 engines. It had a 2,800-square-foot wing with a span of 78 feet 4 inches, was 115 feet 10 inches long, and had a takeoff gross weight of 160,300 pounds. The top speed of this configuration was to be Mach 1.65. (Convair via SDASM)

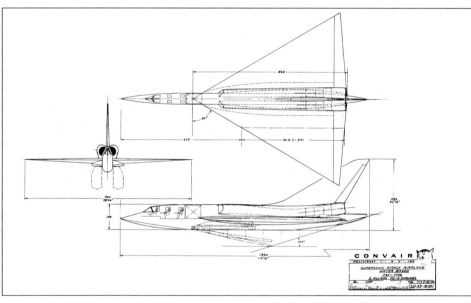

D rawings and a wind tunnel report document a Convair design study of a Supersonic Water Based Attack Airplane in November 1953. Although this work was conducted under Navy funding (NOa[s]-12143), it may have been preliminary to additional work that was reported in June 1955. The summary of the mission requirements was for a Supersonic Attack Seaplane with a combat radius of 1,700 nmi, a 200-nmi sea-level run in and out at Mach 0.9 (minimum of Mach 1.25 at 35,000 feet), carry a 3,000-pound special weapon, and be capable of takeoff and landing in 8-foot

waves. The example covered by this documentation was a twin-tail delta-wing hydro-ski aircraft powered by three Wright J67-W-1 engines. It had a wingspan of 74 feet and a length of 88 feet.

The Navy funded more extensive studies and concept exploration for this mission in 1955 and 1956. The main work was completed in June 1955 and examined several alternative approaches to the mission requirements described above. Two groups of concepts were investigated, the first being delta-wing hydro-ski designs and the second being a swept-wing planing-hull

Seaplane Programs

family of vehicles. The first group included an F2Y-type baseline, several tilting engine concepts, and an engine-in-fuselage concept. The second group included a two-engine version and a four-engine version.

Both existing engines (J75 and J79) and study engines (GE X-84 and Allison 700 PD-9) were considered. The resulting configurations varied in gross weight from 151,000 to 180,000 pounds with the F2Y type in the middle at 160,300 pounds. The deltas all had 2,800-square-foot wings and the swept-wing versions were somewhat smaller at 1,600 and 2,100 square feet. The top speeds were all in the Mach 1.4 to Mach 1.65 range at altitude.

Some continuation of this study resulted in additional work reported in January 1956. Two refined concepts were considered, again a delta-wing ski and a swept-wing planing hull, both with advanced J79 engines. The results of this study indicated the delta-wing ski concept was preferable for several reasons: The delta wing with its large displacement contributed greatly to the aircraft's buoyancy, which allowed a smaller hull, and its rigidity and large internal fuel capacity were desirable features for water-based supersonic airplanes. The conclusion from this study was that a water-based supersonic attack airplane meeting the above requirements was feasible.

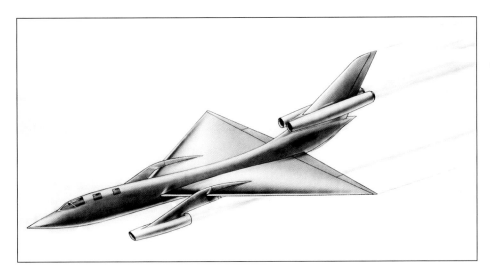

Engine arrangement on the delta-wing SAS may be seen in this flight view. The forward jets were mounted on long pylons to bring them far forward from the center of lift, while the rear engines were tail mounted. Note the heavily area-ruled fuselage of the portion over the wing. (Convair via SDASM)

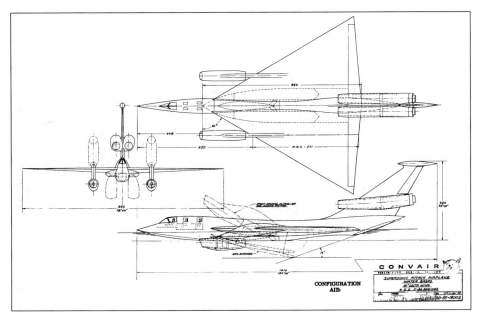

This delta-wing four-engine version features two of its engines mounted on the vertical tail and the other two on pylons on the wing. These wing engines were pivoted so they could rotate upward by 30 degrees in a position for takeoff and landing. These pivoted engines provided a lift vector to shorten the takeoff run and lower the landing speed. In this raised position it also moved the jet intake out of the spray zone. This design was powered by four GE X84 engines, was 134 feet long, and had a 2,800-square-foot wing with a 78-foot span. Gross weight was 162,000 pounds. (Convair via SDASM)

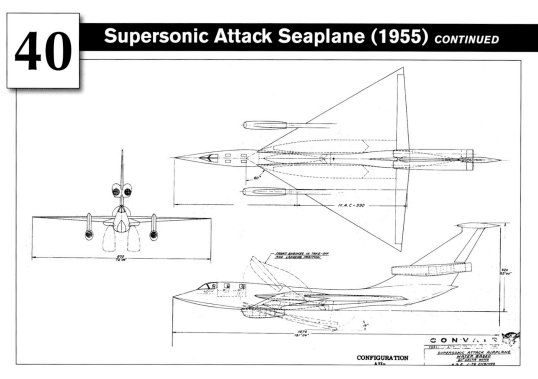

This version has the same tilt-engine arrangement and was similar to the previous version but used GE J79 engines instead. It resulted in a slightly smaller aircraft with a gross weight of 151,800 pounds. In this study a crew of three was utilized for the missions. The nuclear special weapon was carried in a fully submerged centerline fuselage location. *(Convair via SDASM)*

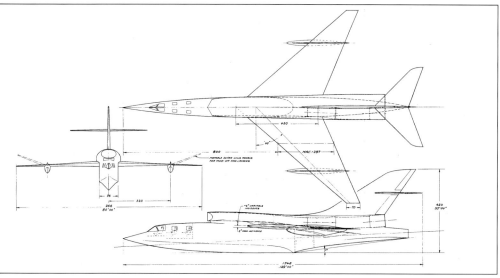

A swept-wing design was also developed for comparison with the delta wings. It had a 45-degree swept wing, a conventional flying-boat planing hull, and wing-mounted floats. The air intakes were located on top of the fuselage as were the two Allison PD9 engines at the far aft position. This aircraft was 129 feet long and had an 80-foot 6-inch wingspan with a gross weight of 161,200 pounds. *(Convair via SDASM)*

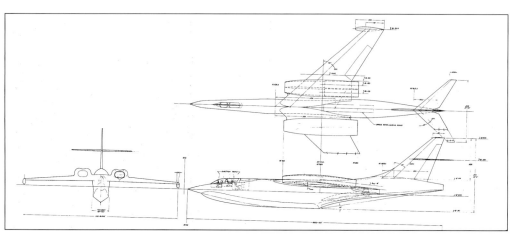

An alternate engine arrangement employed four Pratt & Whitney J75 turbojets mounted in overwing nacelles near the fuselage. The length was about the same as the previous design but wingspan had increased to 91 feet 11 inches. The weapons store appears to be carried in the fuselage bottom just aft of the step in this rare, low-resolution image. *(Convair via SDASM)*

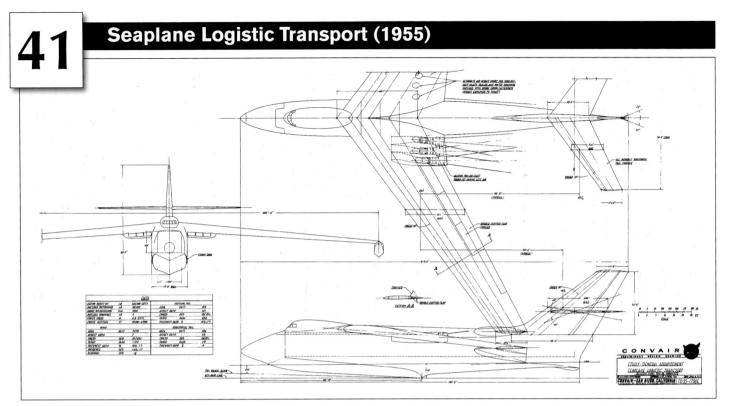

In May 1955, Convair completed a study of a large Seaplane Logistic Transport (SLT) for the Navy that was to evaluate the rapid mobility of troops and heavy equipment in assault operations, and that was optimized for supply cycle time. Two versions were considered, an all-jet configuration and a turboprop-powered version. The all-jet version was powered by six Allison 700 PD engines (maximum thrust 25,000 pounds), had a gross weight of 545,000 pounds, and carried 100,000 pounds of payload. Its 7,270-square-foot swept wing had an impressive span of 225 feet. *(Convair via SDASM)*

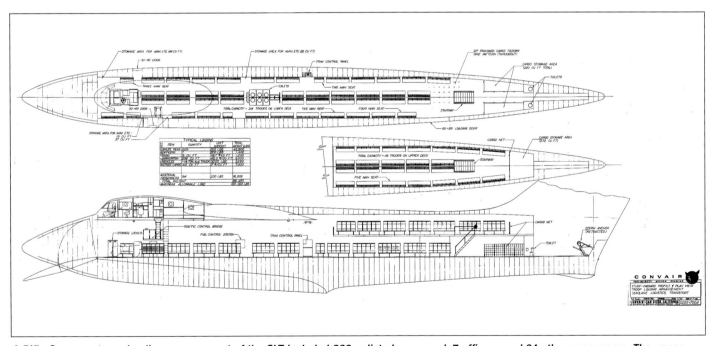

A Rifle Company troop loading arrangement of the SLT included 229 enlisted personnel, 7 officers, and 94 other passengers. The upper deck would accommodate 118 troops with the remainder located on the main cargo deck. The typical loading also included 20,450 pounds of equipment, including a 1¼-ton truck (not shown). *(Convair via SDASM)*

Seaplane Programs

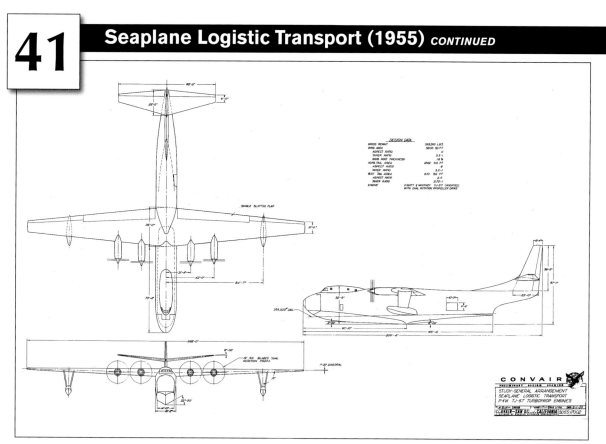

DESIGN DATA

GROSS WEIGHT	393,000 LBS.
WING AREA	5600 SQ FT
ASPECT RATIO	11
TAPER RATIO	3.5:1
WING ROOT THICKNESS	16 %
HORIZ TAIL AREA	842 SQ FT
ASPECT RATIO	6
TAPER RATIO	3.5:1
VERT TAIL AREA	610 SQ FT
ASPECT RATIO	2.5
TAPER RATIO	2.75:1
ENGINE	PRATT & WHITNEY TJ-57 (MODIFIED) WITH DUAL ROTATION PROPELLER DRIVE

CONVAIR
PRELIMINARY DESIGN DRAWING
STUDY-GENERAL ARRANGEMENT
SEAPLANE LOGISTIC TRANSPORT
P & W TJ-57 TURBOPROP ENGINES
CONVAIR—SAN DIEGO, CALIFORNIA SD551002

An alternate SLT was configured that was powered by four Pratt & Whitney XT57-P-1 turboprop engines (15,000 hp maximum). Overall requirements were the same as the jet version including the cargo compartment size and range. This seaplane was somewhat smaller and less complex and had a gross weight of 393,000 pounds compared with 545,000 pounds for the jet version. The straight wing had an area of 5,600 square feet compared with the jet's 7,270 square feet. Range requirement was the same at 1,200-nmi radius, but the turboprop aircraft was significantly slower than the jet versions with a mission average cruise speed of 325 mph versus 460 mph for the jet. (Convair via SDASM)

A study of a Seaplane Logistic Transport was completed in May 1955 aimed at a very large seaplane with a 100,000-pound payload and powered by turbine powerplants. This logistic transport aircraft was oriented to the movement of troop personnel, heavy assault equipment, and cargo into combat or near combat zones as well as non-combat supply areas. The objective of this system was to provide logistic supply with a high degree of mobility optimized for supply cycle time. The requirements included a payload of 100,000 pounds, an unrefueled radius of 1,500 nmi, and a high subsonic cruising speed. It was to have a clear cargo space of 12 x 12 x 80 feet, and be capable of accommodating vehicles and assault craft such as the M-26 tank and surface assault craft. It was to be utilized in conducting tactical operations including unloading at temporary naval bases and at beaches.

Design features included a pressurized fuselage with a wide beam and shallow draft with a large clear cargo space for rapid straight-through loading and unloading. The high L/b and long hull afterbody lines provided efficient aerodynamic and hydrodynamic characteristics and had a wide beam and as minimum a draft as possible. It was designed for rough-water operations at sea with surface craft.

Two large aircraft were studied, a turbojet and a turboprop. The jet-engine version used six Allison 700 PD non-afterburning engines, and would cruise at 460 knots. The turboprop configuration used four Pratt & Whitney XT57-P-1 (PT-5A-1) turboprop engines. It had a takeoff gross weight of 393,000 pounds and a cruise speed of 325 knots with the same payload.

The study concluded that the turbojet option was more complex, had a higher gross weight, and was more costly to operate than the turboprop, but was significantly faster, so the importance of speed could very well have been the deciding factor. Nuclear powered versions of this general configuration were also studied and are discussed in 43. Nuclear Tactical Applications.

Seaplane Programs

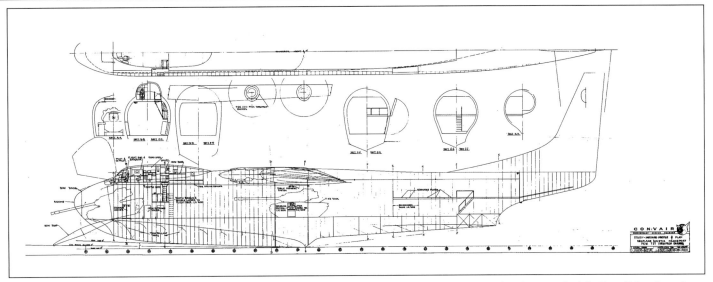

The cargo and flight-deck arrangements were generally similar to the jet version including the optional upper deck in the aft fuselage for personnel or cargo stowage. The upward-opening nose door and the large aft side doors were also similar to the jet version. This particular loading shows the M-26 tank stowed at the mid point of the wing and at the nose door during egress. The profile shows the wing is placed high on the fuselage to maintain adequate propeller clearance from the water. The vertical tail may be seen as uncharacteristically tall and narrow compared to Convair's general design practices. (Convair via SDASM)

42   Nuclear Seaplane - Model 23 (1955)

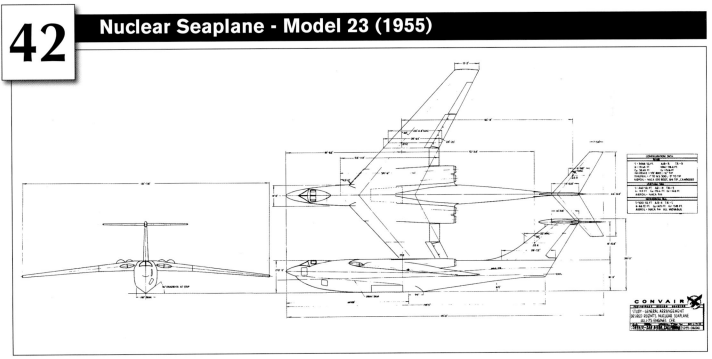

When the Navy first participated actively in the ANP program in May 1953, contracts were given to its seaplane builders, Convair and Martin, for feasibility studies. This then resulted in the Navy issuing an Operational Specification and Development Characteristics for a nuclear-powered seaplane in February and April of 1955, respectively, and contracts were awarded to Convair and Martin. Little technical detail is available on the Convair configuration except for these drawings. This swept-wing design had a span of 131.5 feet and a length of 171 feet. (Convair via SDASM)

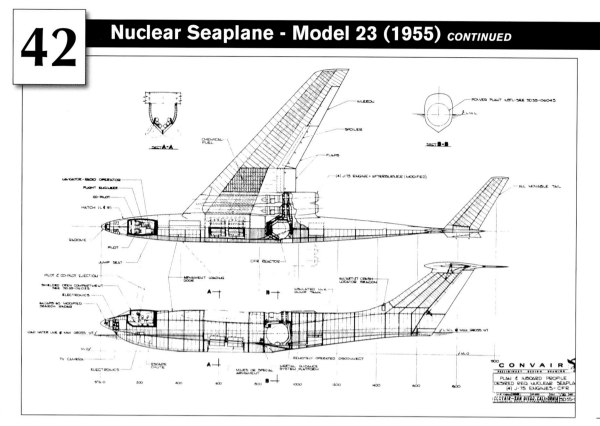

This nuclear-powered aircraft presumably satisfied the aforementioned Navy Development Characteristics. This design was powered by four Pratt & Whitney J75 turbojet engines adapted for operation in conjunction with its circulating fuel reactor (CFR). The CFR was located in the fuselage at the rear of the wing and immediately adjacent to the wing-mounted engines. (Convair via SDASM)

The program for the use of atomic energy for the propulsion of aircraft had a long and turbulent history after its initiation in 1946, during a period when there was widespread optimism with regard to the application and use of nuclear energy. As it turned out, the task of adapting nuclear energy to aircraft propulsion was a very difficult and complex technology problem. A veritable plague of program starts and stops, reorientations and redirections, fragmented management, and a lack of clear direction and guidance only exacerbated these technological problems. On 30 March 1961, the overall program was terminated after an expenditure of $1.04 billion and with a minimum of return realized. A certain amount of reactor and jet-engine development had been accomplished and some reactor facilities had been built and utilized. Convair Fort Worth was the principal airframe participant in the overall program, mainly for a Ground Test Reactor for shielding research and for the NB-36H Shield Test Airplane first flown on 17 September 1955.

Although the Navy was cognizant of and kept abreast of the Aircraft Nuclear Propulsion (ANP) program from its initiation, the first active effort occurred in May 1953. At that time the Navy awarded contracts to seaplane builders (Convair San Diego and Martin) to study nuclear powered aircraft design and provide analyses and data to permit the further assessment of potential objectives of a Naval ANP program. Very little information was available on Convair's work during that initial study period.

The next step was taken by the Navy in February 1955, when it issued an Operational Requirement (CA-01503), and in April 1955 the Development Characteristics (CA-01503-3) for a Nuclear Powered Seaplane. This aircraft was to have a high subsonic capability for long-range attack (primary) and for mine laying and reconnaissance (secondary) missions. Study contracts were again awarded to Convair and Martin to define a nuclear powered subsonic-attack seaplane. A large four-engine seaplane is believed to have been Convair's initial clean sheet design. It was powered by nuclear adapted J75 jet engines and used a Pratt & Whitney circulating-fuel reactor (CFR) for the heat source. Martin is believed to have concentrated on adaptations of its P6M seaplane as a test aircraft during the initial phases of the study. In early 1956, Convair had also briefly looked at additional nuclear designs comparing seaplanes versus WS-110A and WS-125A for the Air Research and Development Command (ARDC).

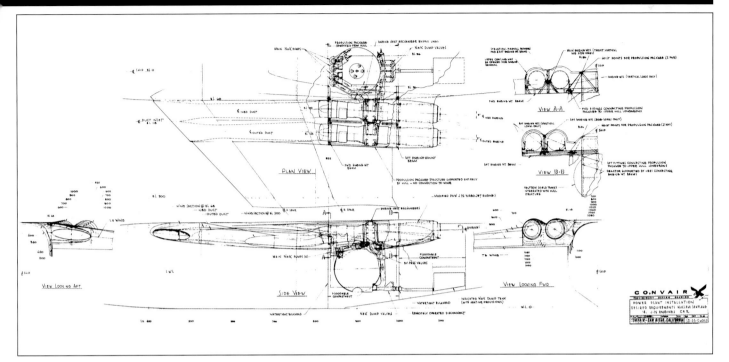

The powerplant system installation of the circulating fuel reactor is shown along with the modified J75 jets. The CFR was an indirect nuclear cycle where there was an intermediate circulating liquid metal (NaK) to transfer the heat energy from the reactor to heat exchangers in the J75 turbojets. The air intakes for the jet engines were in the leading edge of the wing. (Convair via Scott Lowther)

Seaplane Programs

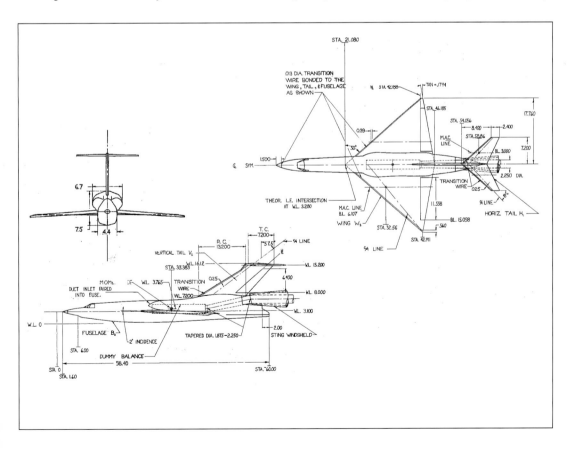

The changing fortunes of the ANP, as well as the Navy's planning impacted Convair study efforts and the configurations were evolving. Wind tunnel testing of the (now) Model 23 designs started in June 1957. This series of designs may have been associated with a research prototype experimental airplane to test nuclear propulsion rather than to develop an operational system. In this particular version the Model 23A incorporated a delta wing as well as a conventional T-tail. The wingspan was 76.3 feet. No further technical details survived. (Convair via SDASM)

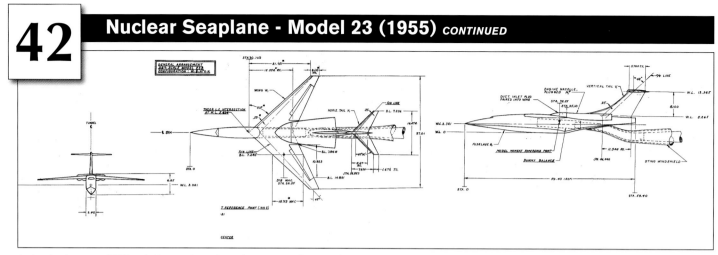

Later, in January 1958, wind tunnel testing of a swept-wing version of the Model 23, the Model 23B, was completed. The fuselage and tail appear quite similar to the Model 23A but a more dramatic 50-degree swept wing was used instead of a delta. The turbojet engines were located over the rear of the wing and the aft fuselage. Wingspan of this configuration was 115 feet. (Convair via SDASM)

Embodying every boy's science-fiction dream in the mid-1950s was this transonic nuclear-powered seaplane known as the Model 23B. Designed for long-range attack missions, aerial mine-laying, and photoreconnaissance, the "Nuclear Seaplane" was simply too advanced for its time, and ultimately unnecessary for the military mission for which it was intended. It was, however, impressively elegant and sleek it was, however. (Extremely rare original Convair factory model of the 23B Nuclear Seaplane and photo courtesy of John Aldaz.)

After a Navy proposal for the development of an alternate propulsion system, believed to be the Allison compact core reactor (CCR), was disapproved by DOD in early 1956, the Navy was directed to use the Direct Air Cycle (DAC) planned for the Air Force's WS-125A program. Many of the design studies Convair conducted during this period used this DAC system concept. By late 1956, however, DOD had lost overall confidence in the progress and prospects of an operational ANP system and canceled the WS-125A. The Navy was considering a turboprop system for its subsonic seaplane mission. In December 1957, the Navy proposed to DOD an early flight-test program for a nuclear turboprop propulsion system using the British Saunders-Roe Princess seaplane. The Air Force was staunchly against the turboprop scheme in the ANP program but said it had no objec-

tions if the Navy supported it with its own funds. In the end the Navy failed to receive DOD approval for the Princess test vehicle project.

In April 1958 the Navy awarded a contract to Pratt & Whitney for additional study of various nuclear systems. In early 1959, Pratt & Whitney received approval from DOD for work on the indirect cycle turboprop system using a lower power reactor than was required by the Air Force program. In December 1959, however, the Navy was directed by DOD to terminate this development program at Pratt & Whitney because it was not in the nation's best interest to have two parallel propulsion systems under development. This ended the Navy participation in the ANP program; however, it was only 15 months before the entire ANP effort was killed in March 1961.

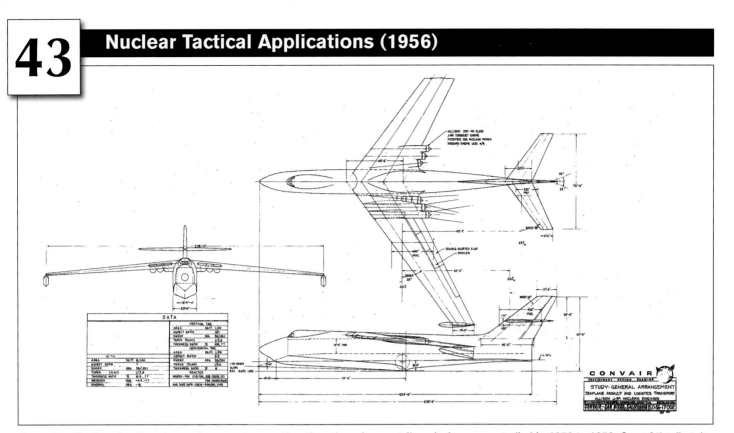

Several tactical applications for nuclear seaplanes, other than the primary strike mission, were studied in 1956 to 1958. One of the first, in August 1956, was a nuclear logistics design. This configuration was a moderately enlarged version of the conventionally powered SLT studied in 1955. The wingspan of the nuclear version was 239 feet, versus 225 feet for the chemically powered airplane, while wing area had been increased from 7,270 to 8,200 square feet. It was to be powered by Allison J89 engines. (Convair via Scott Lowther)

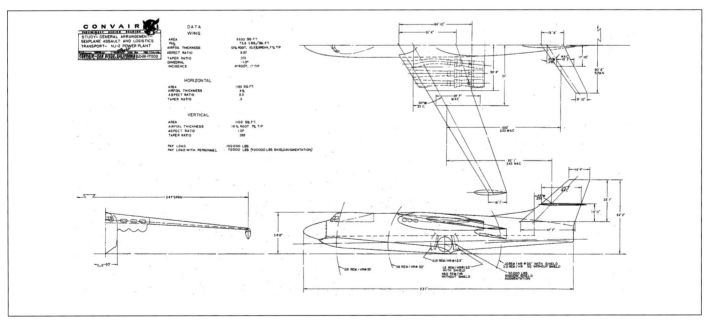

An even larger nuclear logistic aircraft was configured for the NJ-2 powerplant. The reactor was undoubtedly a Pratt & Whitney design, as they were the Navy contractor, but its technical details are not known. This design had a wing area of 8,830 square feet and a span of 247 feet. Payload was 100,000 pounds of cargo and 70,000 pounds for personnel. This latter reduction was necessitated by the additional personnel shielding required around the engines. (Convair via Scott Lowther)

Seaplane Programs

Seaplane Programs

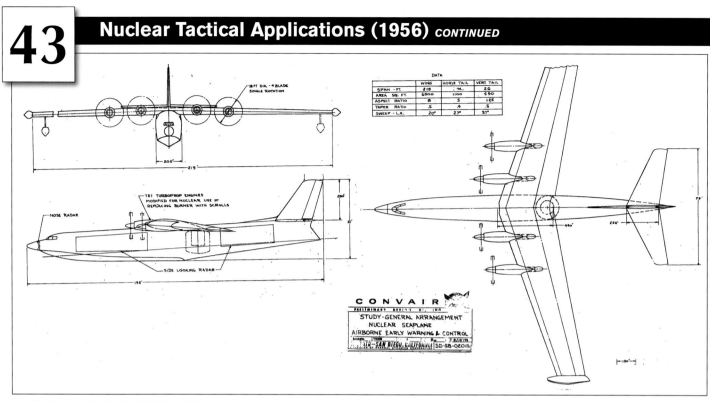

Application studies in the 1957 to 1958 time period were all subsonic and the Navy at that time tended to favor turboprops. An Airborne Early Warning (AEW) application of the nuclear seaplane was studied in April 1957. This design, using moderately swept wings, had a wingspan of 219 feet. The nuclear propulsion system used four T57 turboprops operating off the single CFR reactor. Two very large side-looking radar units on the fore and aft fuselage provided the primary surveillance system. (Convair via Scott Lowther)

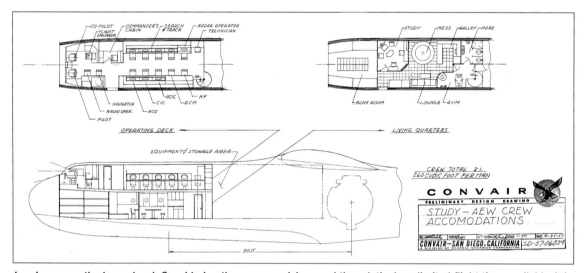

This inboard profile illustrates the crew accommodations on the AEW-mission nuclear seaplane. All of the mission equipment and operator stations were located on the upper level just aft of the flight deck. Crew bunk room, lounge, study, mess, galley, exercise room, and head were on the lower level. Considering the crew provisions and the relatively unlimited flight time available, it is anticipated that very long endurance missions were planned for this nuclear seaplane application. (Convair via Scott Lowther)

Starting in 1956 and continuing into 1959, Convair investigated various applications, in addition to the Navy's primary strike mission, wherein nuclear propulsion could be applied advantageously. Many missions, aircraft configurations, and sizes were considered. Several alternate reactor concepts were also included in the analyses as the Navy's fortunes ebbed and flowed in its attempts to initiate a development program.

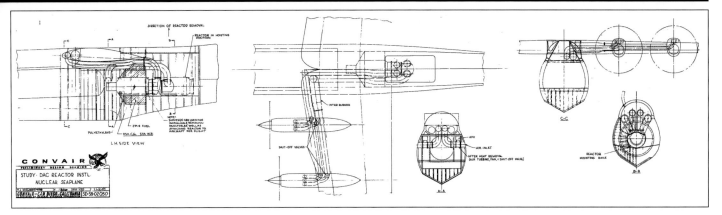

The turboprop engine installation on this aircraft employed a Direct Air Cycle (DAC) engine. Air was heated in the reactor and then pumped to the engine turbine to drive the turboprop. The Navy preferred the indirect cycle system as being better suited to the lesser power requirements of a subsonic aircraft. *(Convair via Scott Lowther)*

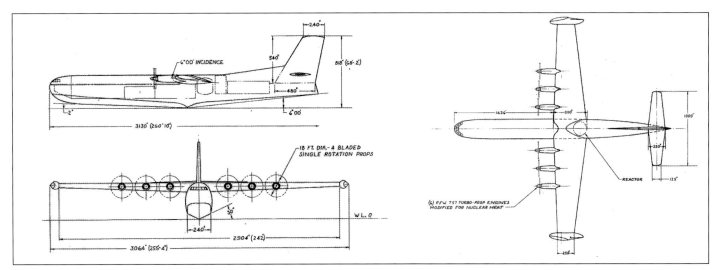

Six Pratt & Whitney T57s powered a logistic aircraft with a gross weight of 750,000 pounds. It had a straight 7,550-square-foot wing with a span of 255 feet 4 inches and the airplane was 260 feet 10 inches long. The single reactor in this design was carried in the fuselage, as in most other configurations. *(Convair via Scott Lowther)*

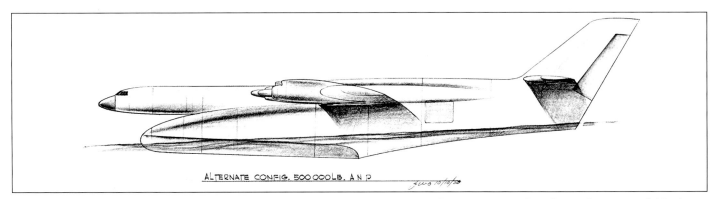

This 1958 drawing of a 500,000-pound aircraft is labeled as an alternative configuration. It appears to be a four-turboprop, straight-wing, nuclear-powered aircraft. The fuselage/hull is reminiscent of some of the early postwar seaplane studies. It is assumed the cargo compartment was in the hull, and the forward-fuselage and flight-deck design assisted in providing radiation isolation. *(SDASM)*

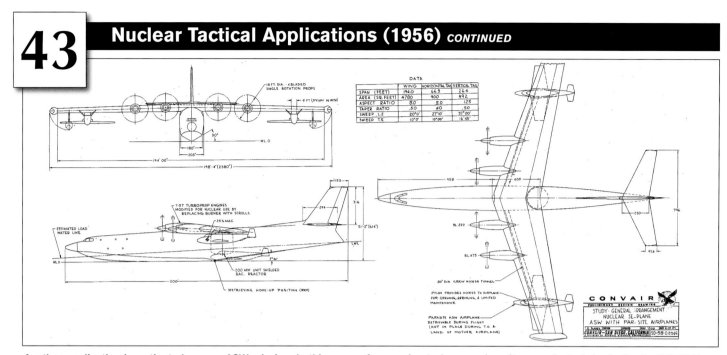

Another application investigated was an ASW mission. In this case a four-engine turboprop aircraft was selected. Again the Pratt & Whitney T57 engines were incorporated in combination with its CFR reactor. The plane had a moderately swept wing with an area of 4,780 square feet and a span of 194 feet. This ASW aircraft carried two piloted parasite and retrievable aircraft for the furtherance of the ASW mission. The operational use for these parasite aircraft is not known. (Convair via Scott Lowther)

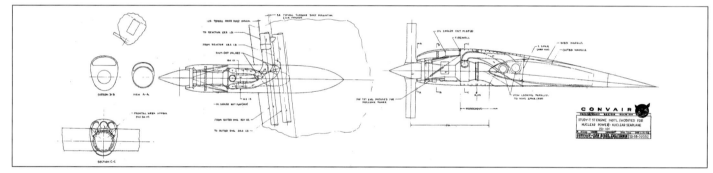

Installation of the nuclear-adapted Pratt & Whitney T57 turboprop engine used with a DAC reactor is illustrated in this drawing. The ducting in the wing leading edge for the reactor's heated air to the inboard engine was 28.5 inches in diameter with 3.5 inches of insulation, and the return duct was 25.5 inches in diameter with 1.25 inches of insulation. The ducting to the outboard engine was correspondingly smaller. (Convair via Scott Lowther)

One of the persistent missions considered throughout these studies was that of the assault transport aimed at moving troops and their equipment into combat areas. A number of size variations of these transports were investigated up to very large aircraft of 1,000,000-pounds gross weight. One of the earliest transports, studied in 1956, employed six Allison J89 engines and was generally similar in appearance to what is believed to be the initial Convair design for a nuclear seaplane, although this application was about twice its size.

Later in the period, other secondary missions were investigated including an airborne early-warning-and-control mission where very long on-station duration was obviously an advantage of nuclear propulsion. An ASW mission was also studied in 1958 and 1959, where long mission times might also be an advantage. All in all these various application studies for the use of nuclear propulsion continued until all Navy ANP work was terminated in late December 1959. Convair San Diego's total funding for its study work on nuclear systems for the Navy totaled $2.9 million.

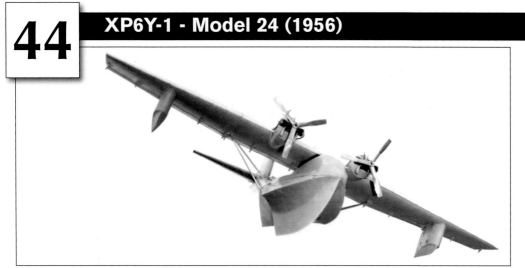

The Navy had developed interest in the concept of deploying dunking sonar from airborne vehicles for the ASW role. After a meeting with the Navy in mid-1954 Convair initiated a study of such a system. The first result was a twin-engine aircraft concept, nicknamed the Dunker, which was extensively tested in the form of a free-flight model. In this low-resolution but rare drawing, note the heavily flared fuselage and deeply sculpted hull, undoubtedly aimed at the open-ocean rough-water problem. This concept eventually evolved to the P6Y design. *(Convair via SDASM)*

This large free-flight model of the Dunker continues Convair's approach of testing models to establish the hydrodynamic characteristics of particular designs. The Dunker was a high-wing twin-engine design with a heavily sculpted hull and a smaller aft fuselage supporting a tail that had a large vertical fin. The wing was mounted in a short pedestal to maintain propeller water clearance. *(Convair via SDASM)*

Seaplane Programs

In the early years of the Cold War it was quite apparent that the Navy's ASW capability would require significant upgrades. One of the technologies showing promise was an approach exploiting dunking sonar for submarine detection, similar to that used on surface ships. In this technique, sonobuoys were dispensed from an open-ocean-capable seaplane that then lands and deploys the dunked sonar-detector unit, receives data, and then takes off to the next location and repeats the process.

Convair had studied an attack ASW seaplane in October 1953 and worked on the problem during the following year. The next noted event was a meeting of BuAer and Convair on the dunking sonar concept on 13 August 1954. Convair proceeded to conduct a study of such ASW operations in October 1954 and developed an aircraft design with two R-3350s, of about 70,000-pounds gross weight, and capable of rough-water landing and takeoff. This initial configuration was visually approaching the design of what was to later become the P6Y. Although no documentation or drawings are available, photos show a large-scale free-flight model of what is believed to be this initial design. This model being tested in late 1955 and early 1956 was nicknamed the "Dunker," undoubtedly derived from the dunking sonar to be used in the operational aircraft. Of note is the widened or flared and very deeply sculpted hull form that was aimed at improving the hydrodynamic characteristics in rough-water operations. The model was also tested with the tilt-float concept although that approach did not appear again until the Post-P6Y ASW studies.

Seaplane Programs

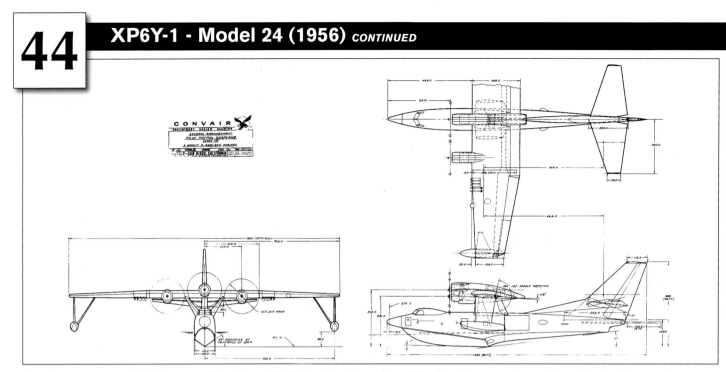

Convair was successful in the competition for an ASW Patrol Seaplane (Class VP) with a three-engine configuration featuring a pylon-mounted high wing in the fashion of the PBY. The engine is a Wright R-3350-32W with a normal rating of 2,800 hp. The propulsion system also included two J85 jet engines that provided full Boundary Layer Control (BLC) flow over the entire wingspan. The P6Y had a span of 127 feet 6 inches and a length of 121 feet. Gross takeoff weight was 107,648 pounds. *(Convair via SDASM)*

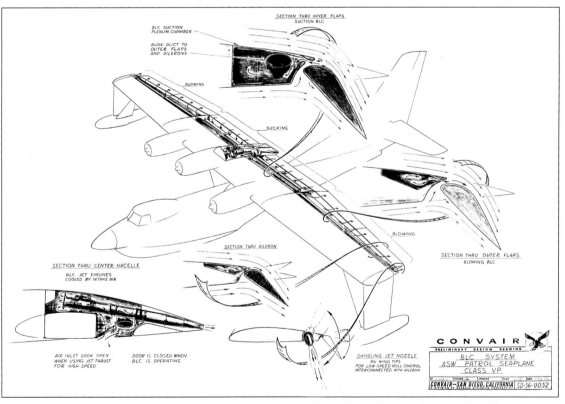

The P6Y incorporated a full BLC system to provide for very low speed controlled flight. The two XJ85-3 jet engines mounted in the aft portion of the center engine nacelle provided both suction and blowing as appropriate to the inner and outer flaps. The inner flaps used suction BLC that pulled air both from the top and the bottom of the flap. The outer flap used a blowing BLC over the top of the flap. Wingtip-mounted swiveling jet nozzles from the same system provided critical low-speed roll control. The P6Y could maintain controlled flight to as slow as 46 knots (power off) and down to an unbelievably slow 40 knots during power-on approaches. *(Convair via SDASM)*

Designed for an ASW role, the tri-motor P6Y was an open-ocean, rough-sea flying boat designed to utilize dunking sonar as a primary detection method. The P6Y's high-mounted wing was supported by a pair of pylons rather than the pedestal and struts of the PBY. Fixed floats were incorporated at the wingtips. (Convair via SDASM)

ASW personnel were on the main deck in the forward fuselage ahead of the wing and behind the elevated flight deck. The large dunking sonar and sonobuoys were housed in a compartment under the wing hull pylons. Immediately aft of the sonar was the weapon stores compartment. The rotating bomb bay doors may be seen in the cross section on the inboard profile. (Convair via SDASM)

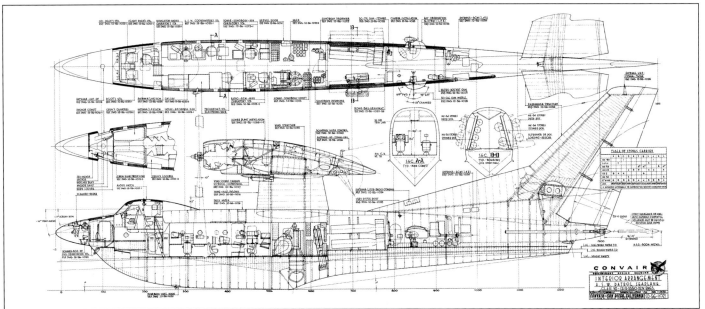

Seaplane Programs

Additional analyses, culminating in January 1956, refined the design with a three-engine concept using Pratt & Whitney R-2800s and a slightly larger aircraft, about 10,000 pounds heavier than the previous design.

The Navy then issued a Request for Proposal for this system on 21 May 1956 and proposals were submitted on 28 August for an ASW seaplane (Class VP). Convair, Grumman, and Martin responded. Martin's design was referred to as the "P7M" (Model 313) and was reportedly a four-engine design using Wright R-1820s. Grumman and Martin were notified they were not the winners in November. Convair then negotiated the detailed airplane specification and submitted revised proposals in July and August 1957. The development and production price for two prototypes was $69.3 million.

Convair's P6Y was an open ocean seaplane utilizing a dunking sonar and was capable of sea keeping ability in 8- to 12-foot-wave conditions. The final design had three Wright R-3350, 2,800-hp engines, a gross weight of 107,648 pounds, and a wingspan of 127 feet 6 inches. A Boundary Layer Control (BLC) system was incorporated that used full-span BLC to permit controlled low-speed flight down to 40 knots for a power-on approach for a rough-water landing. The BLC system was powered by two J85 jet engines mounted immediately behind the outboard piston engines. This

configuration, larger than those previously studied, carried a crew of 10 and
had a range of 3,000 miles.

Experienced seaplane personnel in the Navy were
not enthused with this system and considered the open
ocean rough-water operation to be fairly dangerous as
well as highly uncomfortable. As a result of this lack of
enthusiasm and the fact that the requirement itself was
not that firm, the funds were dropped from the budget
and the Convair P6Y was terminated in December 1957.
Although Convair conducted some further study work,
the seaplane ASW mission was essentially over, being
replaced by the landplane in the form of the P3V Orion.
A unique and unusual configuration related to the P6Y
surfaced in the form of an artist's rendered illustration. It
shows an asymmetrical arrangement of the wing and
fuselage where the main hull is offset with a single pylon
between the center and outboard engines. A single fixed
float mounted on the opposite side of the wing provides
balance on the water. Information was not available that
would indicate the advantages or purpose of this
arrangement and indeed no other documentation or
drawings have been found, possibly indicative of the
eventual judgment regarding this configuration.

This asymmetrically configured P6Y is one of the few such aircraft designs known, at least in the U.S. The Germans designed several such arrangements and historical documents show many such design studies in the World War II period. The motivation for this configuration is not known and no other information is available. (Convair via SDASM)

45 Combat Seaplane (1956)

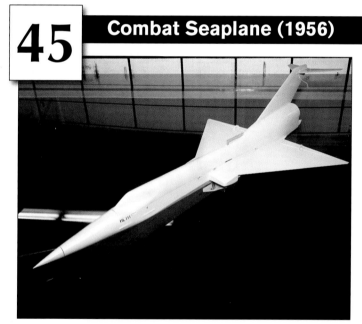

The Navy Combat Seaplane study was accomplished in 1956 and completed in December. The objective was a Mach 3 aircraft to be used for an attack or fighter mission with a combat radius of 800 miles. The study may have been inspired by the previous attack seaplane studies and a proposal for a Navy version of Convair's F-106. This wind tunnel model is one of the variations with a delta wing and conventional tail. (Convair via SDASM)

Convair received funding from the Navy
(NOa[s]56-224C, 3 December 1955) for the predesign of a combat seaplane that was suitable for
an attack or a fighter mission. This aircraft was to be in
the Mach 3 class and have a combat radius of 800 miles.
Attack missions would have a run in and out of 100
miles at Mach 0.9 at sea level or Mach 3 at 60,000 feet
with a 1,700-pound store. This class of aircraft and
mission was suggested by an advanced F-102 study in
1955 and probably Navy attack seaplane studies.

A variety of configurations were investigated
including straight, swept and delta wings, and conventional, canard, and tailless control systems. Landing and
takeoff hull systems included retracting steps, short skis,
and hydrofoils. The propulsion system studies selected
the single Allison 700 B-3 (J89) turbojet as the preferred
engine. Two X-275As were used in a twin-engine configuration as a comparison but were dropped when it
was determined that there were no advantages, probably
a different conclusion than would have been reached by
today's Navy. The airplane would use titanium for the
structure and have provisions for a second crewmember.

Seaplane Programs

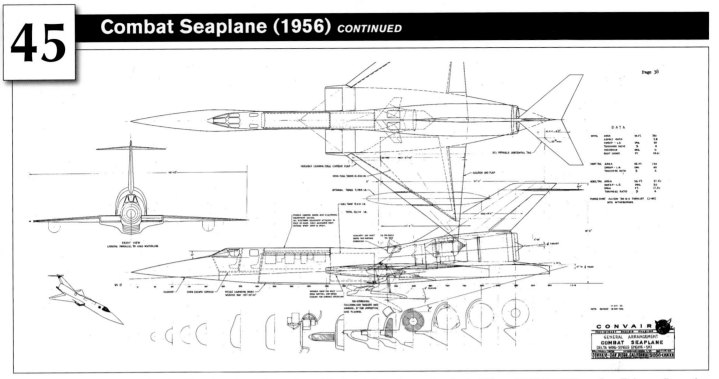

This design variant of the Combat Seaplane two-place aircraft had a delta wing and a single Allison 700B-3 (J89) engine. This configuration had a retractable-ski-hydrofoil for takeoff and landing, a floating fuselage, and a conventional tail for aerodynamic control. The wing had an area of 750 square feet and a span of 46 feet 1 inch. Of special interest are the auxiliary air inlets just in front of the engine on top of the fuselage for use on water, when the main air inlets were closed, to prevent water ingestion. (Convair via SDASM)

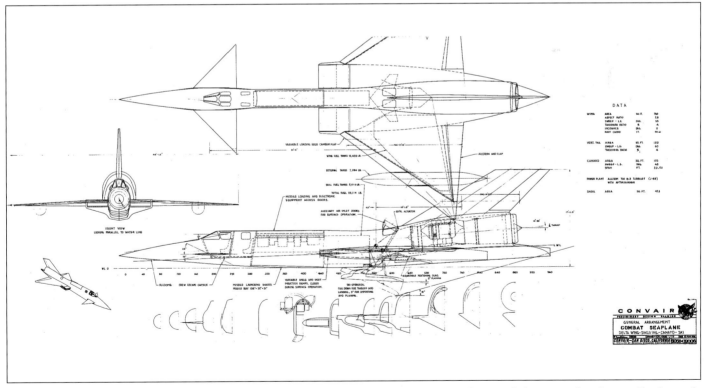

This variation of the initial configuration eliminated the horizontal tail and gained aerodynamic control via a canard surface on the forward fuselage at the cockpit location. Otherwise the configuration is identical. (Convair via SDASM)

Seaplane Programs

Seaplane Programs

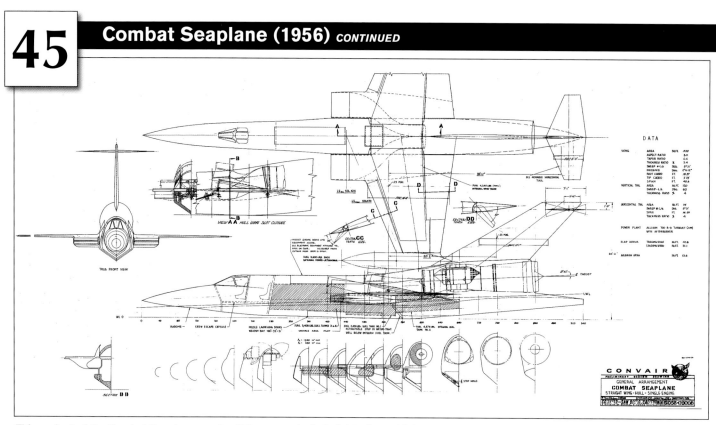

This variant of the Combat Seaplane used a 550-square-foot straight wing, which was somewhat smaller than the delta design, and a hull-type fuselage. The wing on this configuration required variable incidence to attain proper take-off angle of attack in the absence of ski assist. In addition, a retractable aft-fuselage step was incorporated. The disadvantage of the straight wing was its heavier structural weight and lack of fuel capacity requiring increased fuel to be carried inside the fuselage. *(Convair via SDASM)*

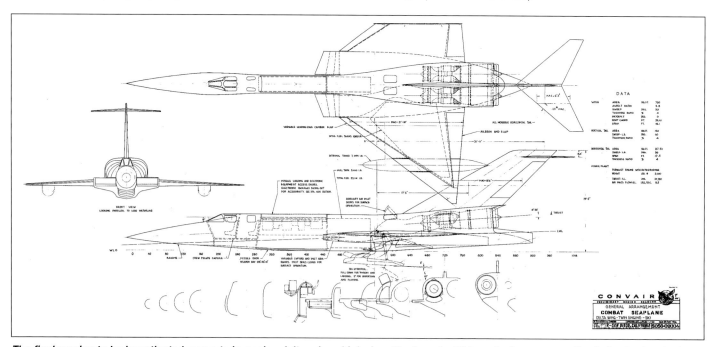

The final version to be investigated was a twin-engine delta-wing ski design. The layout of this variant was generally similar to the single-engine delta ski arrangement. In this airplane two X-275A turbojet engines of 18,500 pounds thrust each occupy the same fuselage location as the single-engine version. This twin-engine approach showed no appreciable advantage over previous designs. *(Convair via SDASM)*

Numerous combinations of the configuration parameters were eliminated and eight configurations were analyzed in more detail. Four of the surviving designs are shown herein.

The first of these was a delta wing, floating fuselage with skis, and a conventional tail. The fuselage had a high fineness ratio and included escape capsules for the crew. The air induction system consisted of a pair of semi-internal compression variable capture area inlets. A secondary inlet system consisted of doors in the upper fuselage for additional air during takeoffs and landings. This airplane with a 2,250-pound payload had a useful load of 28,285 pounds and a gross takeoff weight of 54,335 pounds.

The second configuration was similar to the first design except it had a canard control system instead of the conventional tail. The third design was again similar to the first but incorporated two X-275A engines instead of the single J89. The last configuration used a straight variable incidence wing and no skis on the hull. The wing on this variation was pivoted for takeoff.

It was concluded that either the delta or the straight wing designs could meet the requirements established for the study. As attack airplanes, they would accomplish the 800-mile mission with a Mach 0.9 100-mile run in at sea level, and as a fighter it can reach Mach 3 at over 60,000 feet.

46 Assault Transport Seaplane (1957)

Seaplane Programs

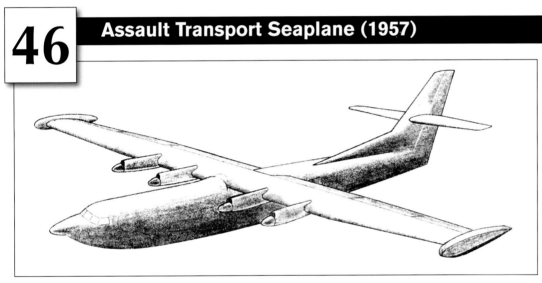

In March 1957 Convair completed a study of an Assault Transport Seaplane. This was a four-turboprop aircraft roughly the same size as the R3Y-2 but significantly heavier at 215,000 pounds versus 145,000 pounds. It featured a variable-incidence wing that would maintain the required propeller/water clearance with a fuselage of lesser height than a fixed-wing aircraft while still maintaining a similar-size cargo compartment. (Convair via SDASM)

Convair initiated a study in early 1957, or possibly in late 1956, for an Assault Transport Seaplane. There is no information to indicate if this was a company-sponsored study or if there was Navy financial support for the effort. The size of the aircraft was roughly that of the bow-loader R3Y-2 Tradewind but had a significantly different fuselage and employed a variable-incidence or tilt-wing design as used in the Supersonic Attack Seaplane study. In this approach the wing could be tilted by up to 17 degrees during takeoff and landing and that then provided a lift force, lessened the hull loads at the water interface, and moved the propellers upward to provide the required water clearance. The latter feature allowed a fuselage that did not require the height of the conventional design.

This transport was powered by four Allison 550 B1 turboprops and could carry a variety of cargo and equipment and accommodate up to 108 troops. There is no documentation available to further indicate the performance capability of this aircraft.

A land conversion configuration was also developed that retained the variable incidence wing for improved takeoff and landing performance. The general appearance of this design can be recognized in the tactical applications of the nuclear seaplanes being studied in the same general time period.

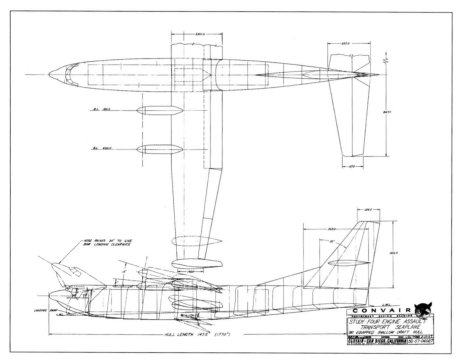

The Assault Transport (above) was designed to deliver troops and equipment to the beach during amphibious operations through a tilt-nose bow loading door. The variable-incidence wing shown here tilted upward 17-degrees and provided a lift force while lessening the loads on the hull as well as maintaining propeller water clearance during takeoff and landing. *(Convair via SDASM)*

The overall size of the Assault Transport was close to the R3Y with a wingspan of 172 feet 6 inches and a length of 147 feet 6 inches. The tilt-nose bow door elevated to 30 degrees to facilitate straight-in cargo loading. This compartment was 9 x 11 x 81 feet 6 inches. The aircraft had a ski-equipped, shallow, draft hull and was powered by four Allison 550 B1 geared turboprops. *(Convair via SDASM)*

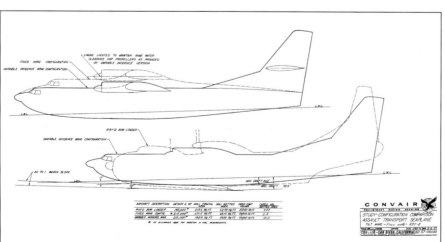

This drawing shows a comparison with the R3Y-2 (top). This Assault Transport shows a significantly increased cargo space of 73,400 cubic feet, almost twice the R3Y's 37,200 cubic feet. The increased payload capability of 50,000 pounds was reflected in the aircraft's significantly increased gross weight. A comparison was also shown (bottom) of the impact of the variable-incidence wing to maintain the same water clearance for the propellers as a fixed-wing configuration. *(Convair via SDASM)*

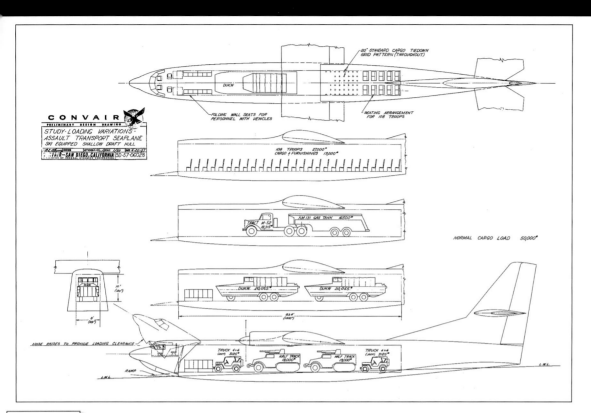

Various loadings of typical items of equipment are shown for the payload bay, with both cargo and personnel. A variety of vehicles could be accommodated, all within the aircraft's normal cargo capability of 50,000 pounds. This aircraft could carry 108 troops, limited by seating but not payload capability. (Convair via SDASM)

<table>
<tr><td>47</td><td>ARDC Seaplane Study (1957)</td></tr>
</table>

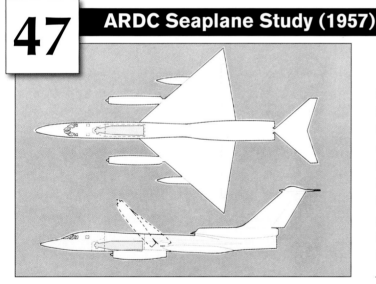

Several interesting configurations emerged during the study Convair conducted for the Air Research and Development Command titled "ARDC Seaplane Study" in 1957. This study was to determine the advantages that seaplanes may have had over landplanes for nuclear weapons delivery. This particular configuration was an equivalent of Convair's B-58 Hustler and appears to have been a version of the Supersonic Attack Seaplane. It had a gross weight of 170,000 pounds, a span of 67 feet 6 inches, a length of 120 feet, and was powered by four GE J79 engines, the same engines used by the B-58. (Convair via SDASM)

In late 1956, the Air Research and Development Command funded Convair to investigate the feasibility of employing seaplanes for the missions then being developed for the four-engine delta-wing Mach-2 B-58 Hustler, the WS-110A that was to become the six-engine delta-wing Mach-3 North American B-70 Valkyrie, and the WS-125A which was the ongoing nuclear aircraft program at the time.

Two configurations were compared with the B-58, a version of the Supersonic Attack Seaplane (SAS) with pivoted engines, and a more conventional seaplane design. No information is available as to the preferred solution.

The WS-110A mission had more difficult requirements for seaplane configurations that led to several unique concepts. Those configurations included a flying wingtip design that was to carry fuel in winged

Seaplane Programs

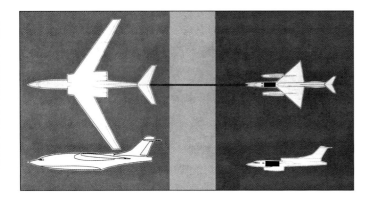

A second more conventional design for comparison with the B-58 is this swept-wing configuration, also powered by four J79 engines. It had a span of 72 feet 6 inches, and a length of 131 feet 8 inches. *(Convair via SDASM)*

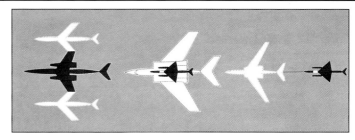

For comparison with the WS-110A (B-70) mission, several concepts were investigated. The first was a design with flying wingtips, a concept that had been investigated extensively during the WS-110A studies. The second was a piggyback concept. The third was a tug-tow concept. The flying wingtip configuration had a combined gross weight of 710,000 pounds, which created serious structural and hydrodynamic problems. The piggyback concept resulted in an excessively large aircraft with a massive gross weight of 1 million pounds. It was then deemed that the tug-tow was the most feasible arrangement. *(Convair via SDASM)*

The unusual tug-tow design used for comparison with the WS-110A involved a large seaplane tow vehicle and a much smaller dash vehicle. The dash vehicle appears to be a pivoting-wing twin-engine SAS type with a gross weight of 160,000 pounds and span of 59 feet 6 inches. The tow plane, with a gross weight of 450,000 pounds and span of 185 feet, would move the combined aircrafts to within the dash plane's combat radius, then disengage and return to base while the dash plane completed the nuclear weapons delivery. (Convair via SDASM)

tanks attached at the aircraft's wingtips. These were jettisoned when the tanks were empty. This floating-wingtip concept was studied extensively in the WS-110A program. The structural and hydrodynamic problems associated with a seaplane using this arrangement were undoubtedly too severe, and it was not studied further.

The next concept was a very large seaplane that carried a second component that would then accomplish the dash mission to the target area, undoubtedly at supersonic speed. As it turned out the carrier for the dash aircraft was prohibitively large with a gross weight of 1 million pounds. The dash aircraft was a two-engine SAS type and weighed 160,000 pounds. This overall concept was also dropped.

The last approach was a tug-tow arrangement where a tug aircraft towed a dash component to within its combat radius, the tug returned to its base and the strike component proceeded to the target. The strike aircraft in this case appeared to be a three-engine version of the SAS pivoting engine approach. Of the three options for the WS-110A mission this one was concluded as the preferred option.

The third and final comparison was with the WS-125A (ANP) nuclear powered landplane bomber studies ongoing during the period. The Convair concepts were, of course, greatly benefited by the Nuclear Seaplane Study being conducted for the Navy. It appeared to have six engines, and the general arrangement was similar to the Model 23B in the Navy study.

No further effort was conducted in this study and it was undoubtedly for information only.

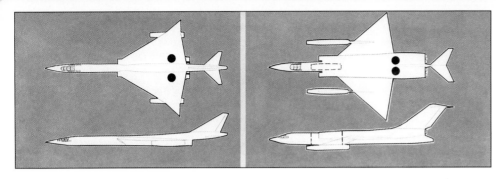

The seaplanes were also compared with the WS-125A nuclear-powered aircraft studies then in progress. A landplane version is shown on the left and a seaplane version on the right. The delta-wing seaplane appears to have four nuclear-powered jet engines and two nuclear reactors plus two conventional turbojets for added thrust during takeoff. *(Convair via SDASM)*

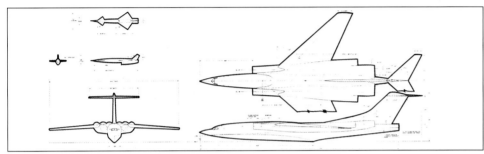

The final WS-125A-type seaplane in this comparison is probably based on one of the Model 23 designs developed in the Navy study. This version had a wingspan of 134 feet and length of 232 feet. Also shown is an air-to-surface missile (ASM) presumably for the final delivery of the nuclear weapon to the target. Its length was 45 feet 5 inches with a span of 17 feet 5 inches. *(Convair via SDASM)*

48 Princess Nuclear Test Bed (1957)

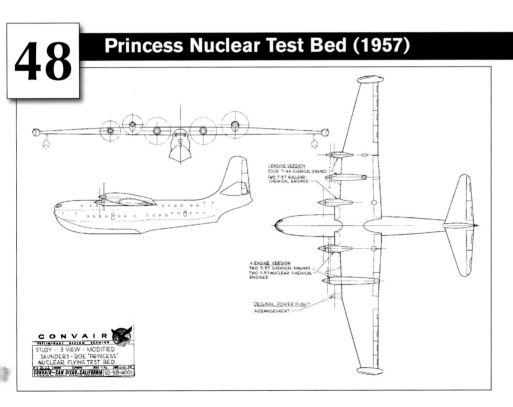

After WS-125A had been canceled in 1957, the Navy strongly considered nuclear turboprops for subsonic missions. In December of that year the Navy proposed to DOD to conduct an early flight demonstration of such a system using a modified Saunders-Roe Princess flying boat that was available. This was a very large seaplane with a wingspan of 219 feet 6 inches. Two designs were considered. The first was a four-engine version with Pratt & Whitney T57 engines: two chemical and two nuclear. The other configuration had six engines: two T57 nuclear and four T34 conventional turboprops. *(Convair via Scott Lowther)*

The Navy had become active in the ANP program in 1953 In 1955, after conducting several studies, DOD directed the Navy to consider the General Electric and Pratt & Whitney powerplants under development for the Air Force's WS 125A.

By December 1956, DOD had lost confidence in the ANP program in that neither the Air Force nor the Navy programs were acceptable. DOD proceeded to cancel the

ACCESS FOR REACTOR CORE REMOVAL

EXTENDED WING L.E.

NUCLEAR-CHEMICAL
P&W T 57 ENGINE

FLIGHT DECK

A-A.

UNIT SHIELD G.E. REACTOR

This inboard profile drawing shows a General Electric Direct Air Cycle (DAC) reactor and shield located in mid-fuselage directly under the wing. This configuration had a relatively short distance to the two nuclear-modified T57 turboprops. The routing of the nuclear-heated air may be seen. (Convair via Scott Lowther)

Air Force WS 125A program. With continued study the Navy was increasingly looking at nuclear turboprops for subsonic missions in that it was claimed to be easier to develop because of the lower reactor power levels required compared to the Air Force supersonic mission. The Navy then proposed to DOD the use of a modified British Saunders-Roe Princess seaplane as a test bed for a turboprop propulsion system tailored to its own mission requirements.

The Princess to be used was one of three experimental aircraft that had been mothballed, and was to be modified by the addition of the nuclear reactor and turboprop engines. The Princess was a large seaplane with a 5,250-square-foot wing with a span of 219 feet

6 inches. The overall length was 148 feet. Convair studied two versions of the test aircraft. The first was powered by four Pratt & Whitney T57 engines; two conventionally fueled and two adapted for use of nuclear heat. The second configuration used two of the nuclear T57s and four T34 chemically fueled turboprops. Several different reactors were studied but the choice seemed to be Pratt & Whitney's indirect cycle system.

The Navy reaffirmed the requirement for the Princess test program to DOD in October 1958 but never received a reply, indicating that DOD did not support the program, and as a result no further work was undertaken on this approach.

49 ASW Studies (1958)

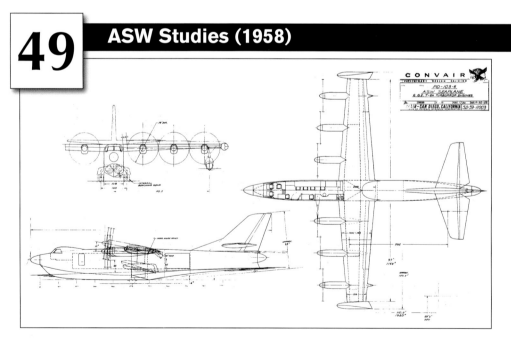

CONVAIR
PRELIMINARY DESIGN DRAWING
PD-103-6
ASW SEAPLANE
6 G.E. T-64 TURBOPROP ENGINES
AIR-SAN DIEGO, CALIFORNIA SD-59-1/003

After the demise of the P6Y, Convair continued to study the ASW problem. The effort concentrated on STOL configurations in early 1959; one of the more promising designs was a six-T64-turboprop configuration with a tilting wing. The wing tilt, up to 30 degrees, provided upward thrust, propeller clearance, and a slipstream effect over the entire wing. These combined to allow a takeoff at 45 knots in only 390 feet. The aircraft's gross weight was 88,700 pounds and the combat radius was 900 nautical miles. (Convair via SDASM)

Seaplane Programs

The P6Y BLC approach also remained of interest during these studies. This version of an ASW seaplane utilized four GE T64 turboprops and a BLC system to obtain STOL performance of a 60-knot landing speed. This was an 86,000-pound-gross-weight airplane with a 1,250-square-foot wing and a 107-foot span. It also used the P6Y rotating bomb bay doors and incorporated retractable wingtip floats. (Convair via SDASM)

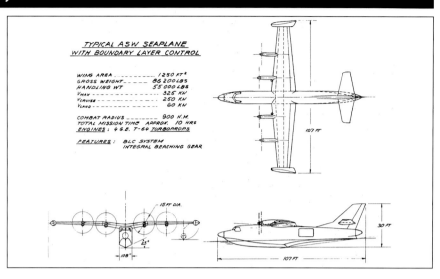

In this typical ASW interior, the flight deck accommodated the pilot, copilot and captain, with the electronics compartment immediately aft. A bunk room was located under the flight deck. The ASW deck compartment had operating positions for the ASW crew, followed by wardroom, sonobuoy compartment, bomb bay, and observer's compartment. The retractable hydro-ski and the retractable beaching gear can be seen in the fuselage beneath the wing. (Convair via SDASM)

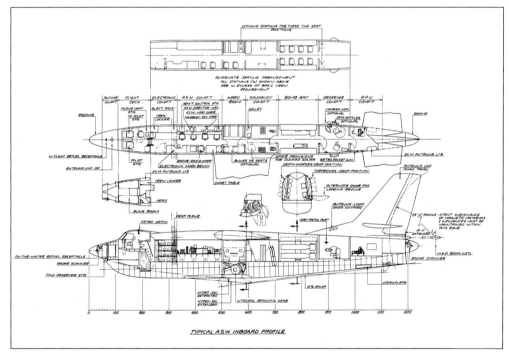

Not long after the P6Y was terminated in December 1957, Lockheed won the competition for a replacement for the land based P2V. The prototype of the new P3V Orion patrol aircraft, a modified Electra, first flew on 25 November 1958 and the first production version flew on 15 April 1961. The trend away from seaplanes and toward landplanes for these missions was well underway.

The Navy's seaplane factions were in the process of rethinking the water-based aircraft and its mission requirements. In April 1958 the Navy issued a new baseline for continuing studies aimed at a 900-nmi combat radius, a cruise speed of 250 knots at 10,000 feet, two hours on station at 150 knots, 1,500 feet, takeoff and landing at 60 knots, a less-than-80,000-pound gross weight, and open-ocean capability.

After the demise of the P6Y, Convair continued study activity and prepared proposals for additional work on seaplanes as well as the overall ASW operational problem. Convair had developed several turboprop concepts by early 1959. This included a BLC design in the manner of the P6Y, and another that was

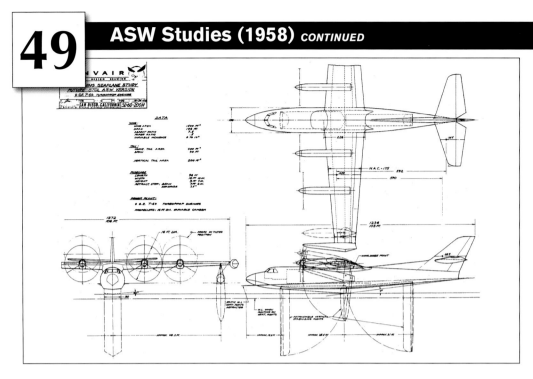

An idea originating with the Navy was the so-called vertical-float method of obtaining stability in heavy seas and was investigated by Convair in this study. In this system, fore and aft retractable fuselage floats about 30 feet long were deployed to the vertical, as were two wingtip stabilizing floats. In this case the bottom of the hull was lifted about 2 feet out of the mean sea level. (Convair via SDASM)

It is believed that the tilt float concept was tested on a P5M. It also appears that Convair was involved as the company was negotiating with the Navy for the use of two non-flying PBM-5G aircraft in January 1961. These test aircraft would be fitted with the appropriate floatation gear and then tested in various sea conditions. The tests were said to have provided a remarkable level of stability. Studies looked at both hard (metallic) retractable floats and at inflatable versions. (Convair via SDASM)

considered of great promise, a six-engine Short Takeoff and Landing (STOL) aircraft with a modestly tilting wing and taking advantage of the slipstream effect. This latter aircraft had a 1,250-square-foot wing with a span of 107 feet, a gross weight of 88,700 pounds, and a wing tilt of 10 to 30 degrees. The six GE T64 turboprops provided a slipstream over most of the wingspan and the upward thrust from the engine tilt provided an unloading effect that greatly reduced the water drag. This effect allowed a takeoff speed of 45 knots, a takeoff distance of 390 feet, and landing speed of 45 to 60 knots depending on the power level.

No information is available as to Convair's contractual study coverage prior to a Non-Planing Seaplane Study (Nas 80-60-6071-c), believed to have started in 1960 and completed at the end of 1961. This study was aimed primarily at ASW STOL concepts although Vertical Takeoff and Landing (VTOL) designs were also considered. A Navy idea that was also investigated closely was the application of a vertical-float concept, also termed a tilt float or sometimes a Spar Buoy float. In this design a retractable or inflatable vertical float was deployed at each end of the fuselage with wingtip stabilizing floats completing the arrangement. This configuration provided excellent roll and longitudinal stability while on rough water. This concept was successfully tested on a PBM-5 but it is not known if Convair was associated with this testing. No documentation occurs in the archives but photographs and illustrations are available.

Seaplane Programs

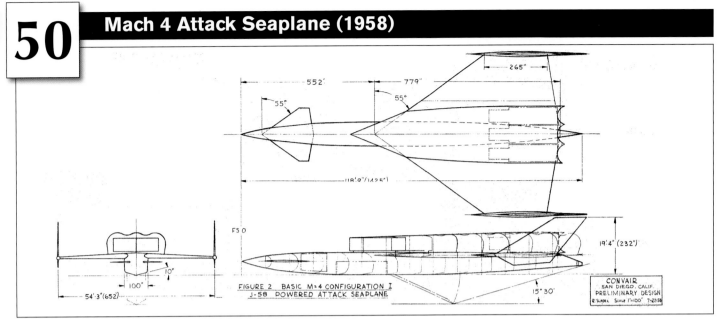

The first Mach 4 Attack Seaplane in the Navy J58 application study was a delta wing canard configuration powered by three Pratt & Whitney J58 turbojet engines. The truncated 55-degree 2,360-square-foot delta wing had a span of 54 feet 3 inches and featured additional vertical stabilizers at the wingtips. The overall aircraft was 118 feet 9 inches long and took off and landed with a hydro-ski. The intakes for the jet engines were located on the top of the fuselage and the crew was located in a capsule buried in the forward fuselage just behind the canard surface. (Convair via SDASM)

Seaplane Programs

In July 1958, Convair completed a preliminary study for a Mach 4, J58-powered attack seaplane. This study was undertaken for the Navy at BuAer's request at a time when their Pratt & Whitney's J58 engine was under development but did not have a firm mission for its use. It is believed that this study was one of several in the industry to identify an application for this engine. In this case the overall requirement was for an attack seaplane that would cruise at Mach 3 at 80,000 feet, carry a 6,000-pound payload, have a combat radius of 1,500 miles, and have a gross weight of less than 200,000 pounds.

Convair's preliminary analysis indicated that Mach 3 was too slow for efficient flight at 80,000 feet so Mach 4 was selected as the design point. The Convair basic configuration was a canard design and a modified delta wing of 2,360 square feet and a 54-foot 4-inch span, and had vertical surfaces at the wingtip. Three J58 engines were located in a single nacelle on the upper surface of the wing. A hydro-ski was used for takeoff and landing. The resulting airplane was sized for 200,000 pounds and a combat radius of 1,675 nmi.

A second configuration was similar in most respects to the basic design but it employed a rollover concept originated at BuAer. In this concept, the aircraft (after takeoff and before landing) rolled over 180 degrees for cruise flight. This left the upper surface of the wing free from nacelles, and placed the inlets, now below, in a more favorable flow field. The escape capsule housing the crew was trunion mounted and rotated as the aircraft rolled. This crew capsule was also elevated during the takeoff and landing process to provide the required visibility.

The third design was a variation on the basic design with an extended engine inlet duct to just aft of the cockpit area. This duct had a translating forward spike for internal compression. This arrangement provided for drag reduction.

Several drawings indicate additional work on a similar concept about one year later, in July 1959. No reports, notes, or other documentation survived so the motivation for this additional work is unknown. The drawings show a twin-engine canard configuration with a highly swept and truncated delta wing. It was a variable-incidence wing that rotated upward, together with the engines, by 16 degrees to facilitate the takeoff and landing conditions. A unique design feature of this aircraft was inclusion of the flight deck capsule for the crew of two that rotated with the variable-incidence wing. This design included a retractable step and end-plate floats at the wingtip.

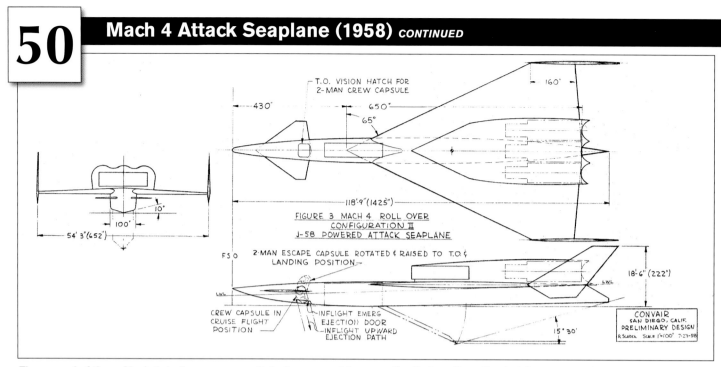

The second of these Mach 4 designs was essentially the same airframe as the first configuration but it was termed a rollover configuration. After the hydro-ski takeoff the aircraft rolled essentially upside down with the engine inlets on the bottom, a more advantageous aerodynamic configuration. The crew capsule was rotated as the aircraft rolled over to maintain a normal attitude. The capsule was also raised during takeoff and landing for improved direct visibility. *(Convair via SDASM)*

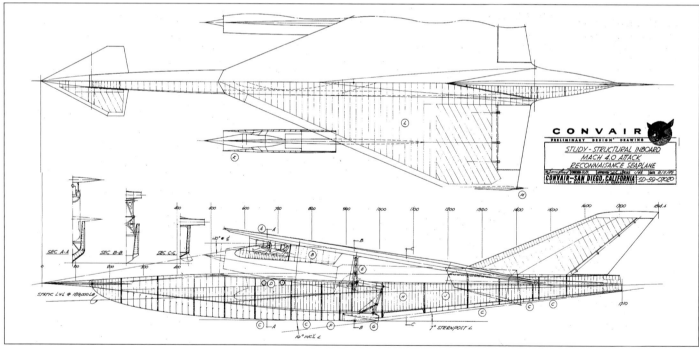

A 1959 design for a Mach 4 Attack Reconnaissance Seaplane is configured around a highly swept delta wing and a canard control surface. The delta wing was a tilt-wing variable-incidence design that moved up 16 degrees for landing and takeoff. The twin engines were pylon mounted on the wing and moved with the wing allowing them to clear the water spray. The the aircraft was 126 feet long and had a 2,640-square-foot wing with a span of 51 feet 4 inches. A unique feature of this design was the crew compartment, which was housed in a manner to rotate with the wing out of the fuselage, providing direct crew visibility for takeoff and landing. *(Convair via SDASM)*

Seaplane Programs

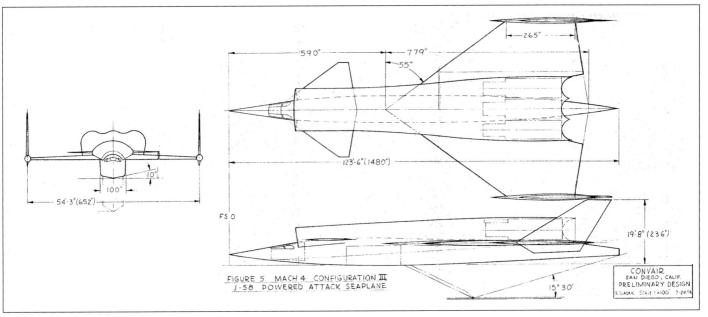

FIGURE 5 MACH 4 CONFIGURATION III
J-58 POWERED ATTACK SEAPLANE

CONVAIR
SAN DIEGO, CALIF.
PRELIMINARY DESIGN

The third design is again similar to the first but used a full-length engine air intake with a forward intake entrance, having aerodynamic advantages with respect to either the over or under (in the rollover design) intakes that were further to the rear of the fuselage. This aircraft was also a hydro-ski and again featured a buried crew capsule. (Convair via SDASM)

51 ASW GETOL (1958)

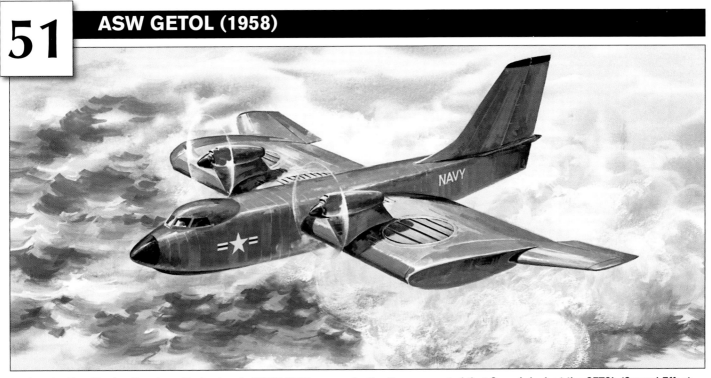

During the Non-Planing Seaplane Study aimed at VTOL/STOL concepts, the Navy requested that Convair look at the GETOL (Ground Effect Take Off and Landing) concept separately. Study activity continued in 1962 and 1963 on a GETOL seaplane for the ASW mission. The general configuration was a very-low-aspect-ratio wing with tip turbine lift fans that provided the capability for ground-effect takeoff and landing. Cruise flight was powered by two tractor-mounted T64 turboprops. (Convair via SDASM)

An alternate design had a more conventional wing, again with tip turbine fans, and utilized a conventional empennage. This version located the cruise turboprop engines in a pusher configuration at the trailing edge of the wing just behind the lift fans. The fuselage and nose were generally similar to the design of the ASW seaplanes, rather than the P6Y-inspired nose treatment. (Convair via SDASM)

A Navy idea also studied by Convair was the tilt-float concept in conjunction with the GETOL study. In this design approach, retractable vertical floats were deployed such that the aircraft was raised above the surface and out of the waves and swells. It was claimed these long floats reaching deep provided excellent stability in rough water. It is believed that the concept was successfully tested on a Martin P5M. (Convair via SDASM)

Convair conducted a comprehensive study starting in the mid 1950s of VTOL and STOL concepts intended to identify areas where Convair could best place its efforts. One of the results of these studies was the identification of an approach termed Ground Effect Takeoff and Landing (GETOL). In this design, ground effects provided by lift fans in the wing facilitated takeoff and landing, and turboprops provided power for cruise flight. The unique low-aspect-ratio wing provided superior weight and structural properties that resulted in improved range and payload performance over that of conventional VTOLs. Studies and wind tunnel work on the GETOL concept were initiated in 1958, both company supported in-house work and contracted research sponsored by the U. S. Army Transportation Research Command. The Navy's Bureau of Weapons (BuWeps) also became interested and provided study money in 1962 and 1963.

The Navy requested that the GETOL concept not be included in the Non-Planing Seaplane study carried out in 1961, but that it be studied separately. An ASW GETOL concept emerged from this work that was in the 80,000 to 90,000-pound gross-weight range with two 100-inch fans in a low-aspect-ratio wing and was powered by turboprops for cruise flight. The typical ASW GETOL would have been sized for a a combat radius of 900-nmi with three hours on station at 1,500 feet and 195 knots. It would have had a payload of 12,000 pounds and a ferry range of 2,600 nmi.

Several variants were studied besides the original low-aspect-ratio GETOL wing, including a more conventional wing and a conventional tail, as well as variations in the location of the turboprops. Most of the designs used two GE T64 engines mounted in either tractor or pusher arrangements and positioned either inboard or outboard of the wing fans. Versions of the GETOL concept also took advantage of the Navy's idea of the tilt-float approach to providing stability on the water surface and to alleviate the concerns with open-ocean rough-water operations and safety.

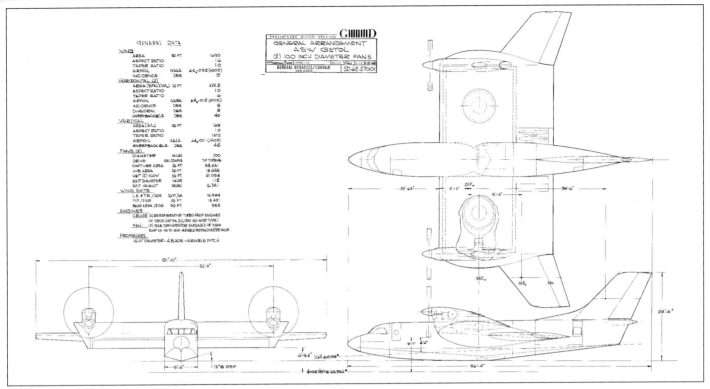

This variant employed the low-aspect-ratio wing, 100-inch tip turbine drive lift fans, and the latter-style fuselage. In this case turboprops were mounted outboard of the fans rather than inboard in the initial configuration. This design had a 1,690-square-foot wing area with an 81-foot-10-inch span and a length of 86 feet 4 inches. Gross weight was 84,000 pounds. (Convair via SDASM)

52 Flying Submersible (1962)

Convair received a small contract from the Navy in early 1963 to study a submersible aircraft that would be used in an ASW role. Reading like something from a science-fiction novel, the overall objective was a system aimed at being able to reach a search area quickly and then conduct the search rapidly and effectively within the target's own environment. A typical mission might have had a 300-nmi radius at 150 knots cruise speed and 10 hours of underwater search at 5 knots and at a depth of 75 feet. (Convair via SDASM)

One of the more unique, if not bizarre, seaplane study efforts engaged in by Convair was a flying submersible proposed to the Navy BuWeps in December 1962. This proposed 34-month study was aimed at an investigation of an aircraft to be used for ASW missions that would be capable of airborne flight, as well as moving submerged in the ocean. Such a system would be aimed at improving the capability to better counter the growing submarine threat. Convair's offer resulted in a six-month study and was funded at

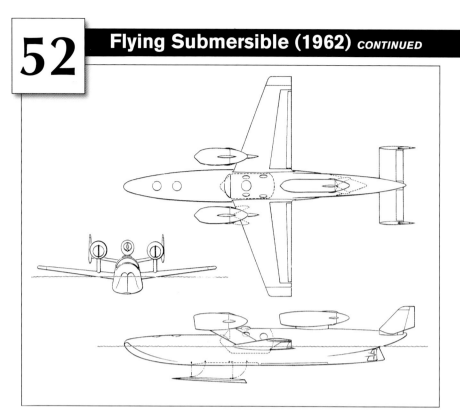

The study was quite abbreviated but did yield a pre-design of a submersible aircraft of the type envisioned. This design, termed a High Density Seaplane, was a medium-size two-place aircraft with a gross weight of 28,000 pounds, a wingspan of about 38 feet, and a length of 52 feet. Payload capability was to be 500 to 1,500 pounds. Three turbojets provided airborne propulsion, and takeoff and landing was accomplished using a hydro-ski. (Convair via SDASM)

Below: This wind tunnel model of the flying submersible shows the overall configuration with little change from the previous drawing except for the forward location of the middle engine and the inclusion of wingtip floats. A small hydro-ski was postulated for takeoff and landings, although the F2Y was to demonstrate that such a design was essentially unworkable. (Convair via SDASM)

$36,000. In addition to analytical and predesign work, both wind tunnel and hydrodynamic model testing was carried out.

Two desirable traits associated with the ASW mission were the ability to proceed to the search area at a high rate of speed and upon arrival to search effectively and quickly. It was generally assumed that the superior approach to the detection and interception of the enemy submarines was to operate in their own environment. A flying submersible was envisioned that would include both of these capabilities.

Convair proposed investigating different VTOL, STOL, and GETOL approaches to make open-ocean operation feasible for the aircraft aspect of the vehicle, and various underwater propulsion systems including conventional propeller, pump jet, cyclic underwater gliding, and buoyancy propulsion for the underwater phase of operations. A typical mission requirement would include an airborne 300- to 500-nmi radius at 150 to 250 knots and an underwater search of 10 hours of operation at 5 knots, and at a depth of 75 feet.

One of the resulting concepts was termed a High Density Seaplane of about 28,000-pounds gross weight, a wingspan of about 38 feet, a length of 52 feet, and carrying a crew of two. It was to be capable of carrying a payload of between 500 and 1,500 pounds. Three tur-

bojets, one for cruise and two additional for takeoff, provided the airborne propulsion. A hydro-ski was used to facilitate takeoff and landing. For the underwater phase all non-critical areas would be flooded to eliminate large pressure vessels.

At the end of this short study Convair felt that such an aircraft was feasible but there was apparently no further Navy interest.

Ending nearly six decades of proud military service with ocean-going flying boats, including numerous Consolidated and Convair types, the Martin PBM Mariner finally brought down the curtain on flying-boat operations for the United States Navy in July 1968. (National Archives via Dennis R. Jenkins)

Indeed the Navy's supercarriers and landplanes prevailed and virtually all seaplane development activities ceased by the early 1960s. Carrier-based strike aircraft like the A2D and A3J and, of course, the Polaris ballistic missile had satisfied most of the Navy's attack mission requirements. The Navy had developed interest briefly in nuclear-powered seaplanes, primarily as a competitive effort at a time when the Air Force was aiming to preempt that capability. But the seaplane's fate was already sealed. There were some low-level study efforts

of the use of seaplanes for ASW missions that continued for several years, including the initiation of the short-lived abortive P6Y program. Seaplane missions were confined to ASW and even those evaporated entirely soon thereafter in the face of carrier-based and land-plane and missile capability. The majority of all Navy seaplane operational activity ceased with the disbanding of the VP-40 squadron of Martin P5Ms in 1968, signaling the end of an era. The last official SP-5B flight took place on 8 July 1968.

CHAPTER THREE

BOMBER PROGRAMS

In development before the end of World War II as this nation's largest long-range strategic bomber, the XB-36 set new records for payload and endurance. Shown here being trailed by a Consolidated C-87, military transport version of the famed B-24 bomber, the XB-36 takes to the skies as the final result of many concept studies that ranged from flying-wing designs to twin-tailed six-engine "hemisphere bombers." Eventually named the Peacekeeper, the B-36 became best known as the airplane that helped win the Cold War without ever firing a shot in anger. (National Archives via Dennis R. Jenkins)

Consolidated, from its inception in 1923, had developed quite a successful trainer airplane business, but by the late 1920s and early 1930s, the beginning of the industry doldrums years, Reuben Fleet was looking to expand Consolidated's business capability into other aviation markets, namely larger multi-engine aircraft. After an initial failure, a joint venture with Sikorsky for the Guardian heavy bomber, Consolidated seemed confined to its very successful seaplane market. In the late 1930s there was a burst of design studies for Army bombers but no developments ensued. However, as World War II loomed, the B-24 was born and Consolidated had another blockbuster winner. The following is a review of the landplane bombers that were studied and proposed, and with reference to those, actually built at the Consolidated/Convair San Diego location.

In 1928, Consolidated joined with Sikorsky to propose Sikorsky's S-37B Guardian twin-engine biplane in the Army's Night Heavy Bomber competition. Sikorsky would build the prototype and Consolidated would then produce the planes. The bomber had a wingspan of 100 feet, a gross weight of 14,650 pounds, and a crew of five. The proposal was unsuccessful in the competition because the aircraft was overweight and had poor performance. (Convair via SDASM)

The Army, in the 1926/1927 time period, had been funded to re-equip its bomber forces. After conducting several experimental bomber projects, the Army initiated a competition for a new Night Heavy Bomber in 1927. Consolidated had been looking into market opportunities for larger and multi-engine airplanes and took the step of teaming with Sikorsky on this competition. Under this April 1927 agreement, Sikorsky would build the prototype airplane for the competition and Consolidated would then be responsible for the production since Sikorsky did not possess sufficient facilities. The entry, named the Guardian (Sikorsky Model S-37B and CAC Model 11), was a twin-engine biplane aircraft adapted from Sikorsky's S-37 built for Captain Rene Fonck a year earlier. It was powered by two Pratt & Whitney Hornet engines of 525 hp each, had a wingspan of 100 feet, and had a gross weight of 14,650 pounds including a bomb load of

1,952 pounds. Its maximum speed was 128 mph. It carried a crew of five: pilot, bombardier, relief pilot, and three gunners, one of whom was the radio operator. The prototype was completed on 9 December 1927 and flew shortly thereafter. It was given the Army bailment designation XB-936.

Unfortunately this project turned out to be an also-ran in the Army's competition because of design deficiencies, including being 1,000 pounds overweight, and poor performance. Although the Curtiss XB-2 had the best performance, in February 1928 the Army selected the smaller and cheaper Keystone XLB-6 as the winner. Curtiss was, however, awarded a contract for 12 B-2s later in June. That experience was a disappointment to Maj. Fleet and was the first and last joint venture he engaged in. It also appears to have been Consolidated's last activity for large Army aircraft for almost 10 years.

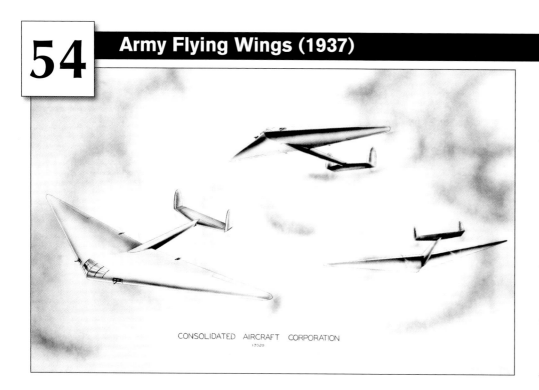

CONSOLIDATED AIRCRAFT CORPORATION

Consolidated investigated what was termed a "flying wing" in late 1937. Although not a true flying wing as we know it today, it did include the engines, fuel, crew, and payload within the wing structure. Many versions of this configuration were investigated including bombers, cargo and passenger transports, and seaplanes. The basic design located the crew in the leading edge of the wing, had twin pusher engines buried inside the wing, and a slim aft fuselage boom supporting twin tails. (Convair via SDASM)

The earliest bomber studies that were found in the archives were the so-called flying wing studies carried out in 1937 and into 1938. As pointed out in Chapter Two, the configuration of these designs was not a wing-only design that is known today since it did include a conventional twin-tail empennage supported by a very small fuselage structure. All of the crew, payload or cargo, passengers, equipment, engines, and fuel were, however, accommodated in the wing itself.

A wide variety of applications for this flying wing were investigated, including bombers, transports, commercial passenger versions, and some seaplane applications discussed earlier. All of the initial versions of this design centered around a wing that had been subject to wind tunnel testing in both the basic configuration and with underwing fairings for increased payload capacity.

The initial airplane configuration in this study was a twin-engine bomber (LB-37-0015 and LB-38-0001) with buried Allison XV-3420-1 liquid-cooled engines. Gross weight of this airplane was 50,000 pounds and it had a range of 4,520 smi with 2,000 pounds of bombs that were carried externally under the wing center section. The maximum speed was 321 mph. The defensive armament for this bomber included two retractable fuselage turrets and two forward flexible guns in the wing leading edge between the flight deck and the engines. Each of these gun positions included a single

37mm cannon. It appears the aircraft had a crew of five to seven located in the wing leading edge in the center section. In addition to being a bomber, this airplane was also configured for a ground attack mission with its armament including eight .50-cal. and two .30-cal. forward-firing guns mounted in the wing leading edge.

A cargo transport version was studied that included a modification of the basic configuration by the addition of a fairing below the wing to provide payload bay space that was not available in the wing itself. Two versions were envisioned, a shallow fairing and a deeper one that would accommodate a larger volume of payload.

Several commercial passenger transports were studied based on the basic twin-engine design. In these applications it was chosen to use the under wing fairing to accommodate the passengers. The first of these configurations was a 40-passenger airplane. Passenger seating was arranged in the shallow version of the fairing on either side of the center section but inboard of the engines. Other versions utilized the deep fairing but still retained the seating at 40 passengers.

Recognizing the limitations of the original flying wing and the necessity of a somewhat awkward under wing fairing, the conventional fuselage then started to reappear somewhat later in the conceptual designs. Both a twin-engine 40-passenger and a larger four-engine 100-passenger commercial transport were studied where the design reverted to a conventional fuselage but

Bomber Programs

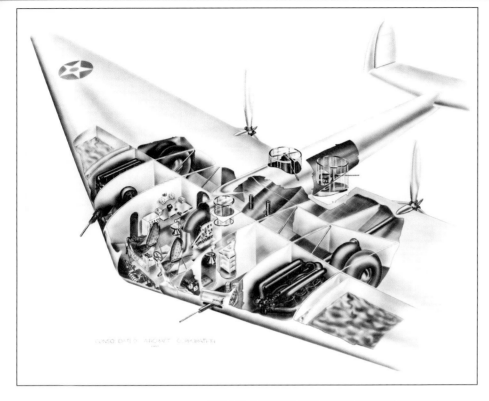

The engine installation driving the pusher propellers is shown in this artist rendition. Flight deck in the nose appears to accommodate five of the aircraft's crew. Two retractable gun turrets were located on the aft fuselage and two flexible guns were mounted on the wing leading edge on either side of the flight deck. Large nose wheel retracted into the aft center of the flight deck. (Convair via SDASM)

Cockpit flight deck illustrates good forward and upward visibility. The bombsight (and a knee pad for the bombardier) was to the left of the pilot's seat. To the right of the co-pilot was a camera, a viewfinder, and a stool for the operator. Main strut of the nose landing gear is between the pilots' seats. (Convair via SDASM)

retained the general idea of the flying wing. The passengers were still accommodated in the wing area with a few seats in the forward part of the aft fuselage. The flight deck was located forward in a rather short nose section. This design was essentially the same 50,000-pound airplane as the previous designs; it had the same 110-foot wingspan, and the same pusher engines. It had a range of 2,370 smi and a top speed of 349 mph.

An enlarged 100-passenger version was also studied that scaled up the basic planform for a four-engine airplane. This design had a 3,200-square-foot wing with a span of 180 feet, a length of 86 feet, and a gross weight of 104,000 pounds. It carried the passengers on two decks; 80 on the lower deck in the wing as before and 20 on the upper deck, mainly in the fuselage area. The range of this airplane was 2,560 smi, similar to the two-engine version, but the top speed at 248 mph was significantly slower.

By mid 1938 the flying wing as initially conceived, as well as its design evolution, had been abandoned.

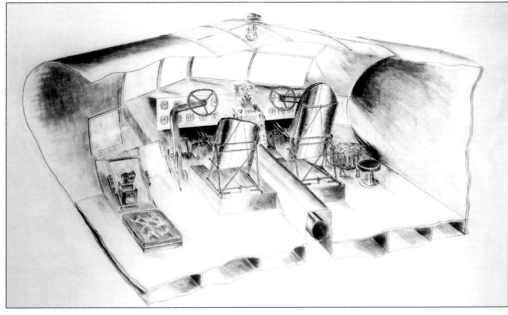

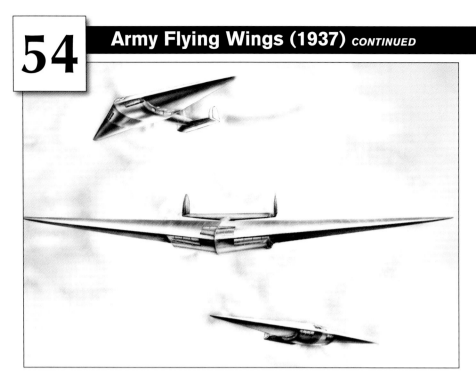

The basic design, initially configured as an Army bomber, was soon modified for the transport of cargo and passengers, and as a seaplane. A shallow under-fairing was developed for passenger versions of the flying wing. This version had the forward windows and some side windows for passengers. Again, as with the deep fairing design, the remainder of the aircraft was essentially unchanged. *(Convair via SDASM)*

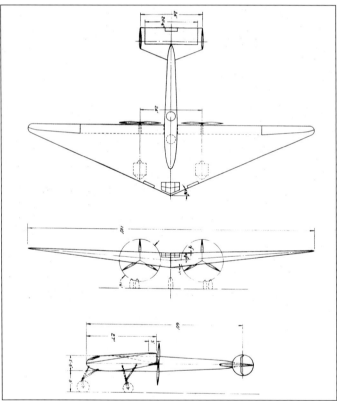

Basic configuration considered at the beginning of the study had a 1,650-square-foot wing with 110-foot span. The airplane's gross weight was 50,000 pounds, and two Allison V-3420-1 engines of 2,000-hp each installed in a pusher configuration powered it. Range was 4,000 miles with 4,000 pounds of bombs and top speed was 321 mph. *(Convair via SDASM)*

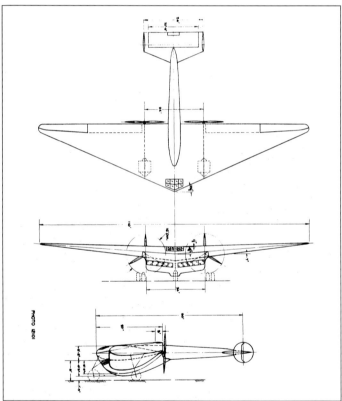

This transport version used the deep-fairing modification of the basic configuration. This drawing is identified as one of the 40-passenger airplanes because of the forward windows in the under-fairing. The twin tail had, at this point, evolved to a mid-surface mounting on the stabilizer rather than the earlier bottom attachment, undoubtedly for structural reasons. *(Convair via SDASM)*

Bomber Programs

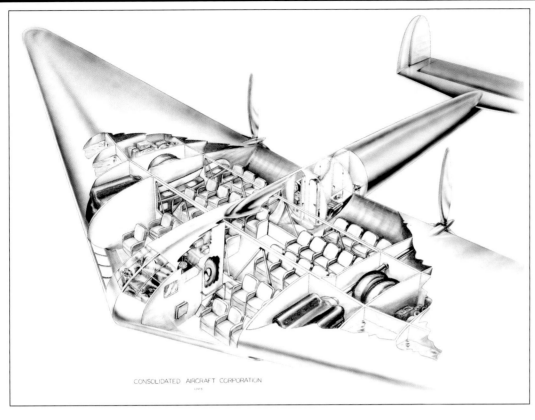

CONSOLIDATED AIRCRAFT CORPORATION

Seating for the 40-passenger design is shown in this perspective inboard profile albeit with a few problems in terms of passenger comfort. Front rows were tiered "theater" seating while the ones to the rear were all on the same level and perhaps a bit claustrophobic for passengers. While the front seats had excellent views via the front windows, they were in close proximity to the engines, which certainly would have required ample sound insulation. Lavatories were in the center fuselage aft of the seating area and the baggage compartments were outboard of the main wheel wells. (Convair via SDASM)

CONSOLIDATED AIRCRAFT CORPORATION

Interior view of the passenger cabin shows some wing structure between the forward and rear parts of the cabin. Forward seats would have an excellent view, weather permitting. The enclosed flight deck is shown at the left of this rendition and there appears to be provisions for passenger refreshments at the back of the crew compartment. It is not clear whether there was passenger movement from the front section to the rear. (Convair via SDASM)

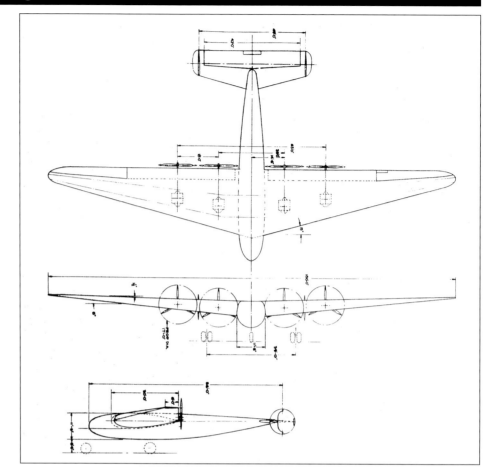

A large passenger version of the basic planform was studied; in this case a four-engine airplane accommodating 100 passengers. This design had a wingspan of 180 feet and a length of 86 feet. Passengers were located mainly in the wing area as with the smaller 40-passenger aircraft, with some seated on an upper deck. *(Convair via SDASM)*

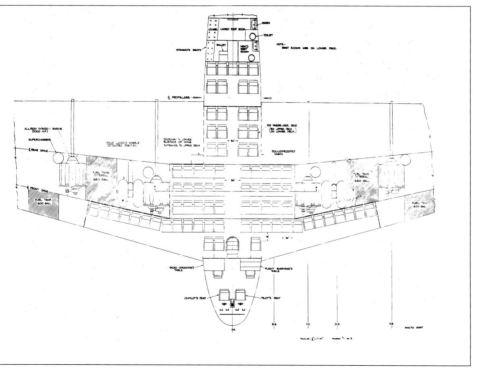

Layout of the passenger compartment on the 100-passenger airplane was generally similar to the smaller transport with a main area in the center section of the wing. About 20 seats were located in the wing leading edge (with a view) and about 20 in the fuselage at the same level as the main compartment. In addition to the main deck, the remaining seats were presumably located in the lower deck. (Convair via SDASM)

Bomber Programs

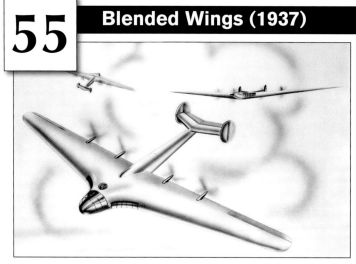

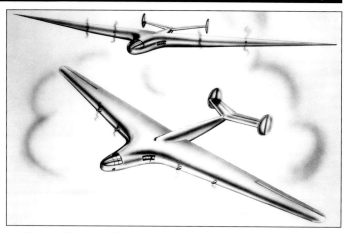

As these drawing-board designs evolved, the configurations again acquired a full and functional fuselage in recognition of the limitations of the flying wings. This four-engine bomber shows a clear legacy from the previous flying wing design with buried pusher engines, twin tail design, and wing planform. None of the documentation or design data survived so no other details are available. (Convair via SDASM)

In March 1938, a tractor version of this bomber was investigated that appears to be virtually identical to the pusher design except for the engine installation. It also had a slightly longer nose, possibly due to the shift in the center of gravity caused by the change in engine location. (Convair via SDASM)

This illustration of the interior of the bomber shows the flight deck in a rather short nose followed by a compartment with the support crew. The fuselage bomb bay was immediately aft of the support-crew compartment. Defensive armament included two retractable turrets top and bottom, a gun position on each side of the forward wing root, and two side positions at the mid aft-fuselage. Turret and forward guns may have been 37mm as in the flying wings. (Convair via SDASM)

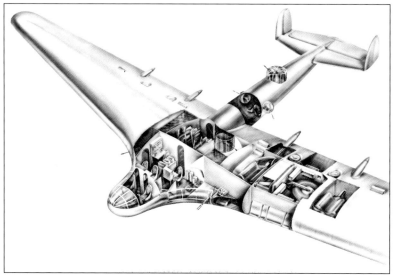

As the designs changed away from the flying wing concept, the use of a full fuselage returned to better accommodate the crew and payload, and either military equipment or passengers. Another trend of this period was the heavy use of aerodynamic blending between the wing and the fuselage. This is seen in both the bomber aircraft design and in the seaplanes discussed earlier. These two bomber designs from March 1938 are noted in the photo logs as being in response to Army Type Specification 98-208. No other information concerning this requirement is known.

One of these follow-on designs was a four-engine twin-tail bomber that shows a clear next step away from the flying wing. It is believed that, because of the buried pusher installation, it used the Allison XV-3420 liquid-cooled engine that was incorporated in many design studies of the period and had been used in the earlier flying wing. This design had a short nose, heavily glassed flight deck, and crew stations at the blended area of the wing root. This latter crew station included flexible forward-firing guns. There were two retractable turrets similar to the flying-wing installation, one dorsal and one ventral, and what appear to be two flexible guns in the fuselage waist section, one on each side.

In addition to the pusher-engine configuration, a tractor version was also studied that appears to be virtually identical except for the engine installation.

Bomber Programs

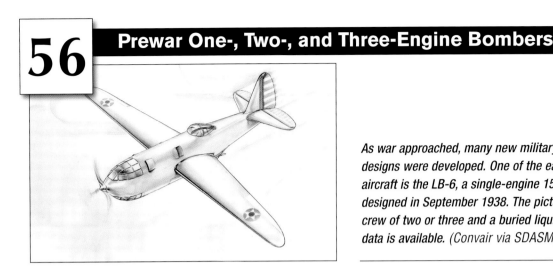

As war approached, many new military attack and bomber aircraft designs were developed. One of the earliest images of landplane attack aircraft is the LB-6, a single-engine 15,000-pound attack bomber designed in September 1938. The pictured airplane appears to have a crew of two or three and a buried liquid-cooled engine. No dimensional data is available. (Convair via SDASM)

As the threat of a major war began to emerge in the late 1930s many in the military and the government began laying the groundwork for a more prepared America, but in general it was an uphill battle. Maj. Gen. Henry H. "Hap" Arnold had been appointed to head the Army Air Corps (AAC) in September 1935 and was a long time airpower champion. He had fostered the Boeing B-17 long-range bomber development program that made its first flight that same year. Soon after Hitler occupied Austria in March 1938, President Roosevelt called for the production of 10,000 airplanes per year as a statement that the United States was not going to be unprepared. After Hitler invaded Poland in early September 1939, the British declared war. After a relatively quiet period of about eight months, Hitler moved through Western Europe, initiated the Battle of Britain, and World War II was at full force in Europe. Although full-scale mobilization was strongly resisted in the United States, in May 1940, Roosevelt again called for the production of 50,000 new airplanes. Planning was getting underway, and by the end of 1941, after the attack on Pearl Harbor by the Japanese in December, mobilization was on a full-scale basis.

The Air Corps, in 1937 to 1940, was engaged in many projects and issued many requirements searching for aircraft to meet the varied needs of trainers and fighters as well as attack, bomber, and transport aircraft. After several years during the doldrums of Depression recovery, the predesign departments of the military aviation companies were working at top speed.

Consolidated had considered the somewhat radical flying-wing design during 1937 and 1938, but the design underwent a continual evolution toward a more and more conventional configuration. Consolidated abandoned that specific concept by early 1938, when it likely failed to generate sufficient interest with the customer. During this period, late 1937 and 1938, many design studies and proposals were undertaken concerning attack and bomber aircraft. Most of these, if not all, seemed to be in response to specific requirements of the Army Air Corps. The studies undertaken during this two-year period culminated in the B-24 and the B-32 programs. There is not enough documentation in the Consolidated archives to trace many of the actual requirements and the evolutionary sequence of designs in the various programs.

The predesign designation system for these advanced designs used by Consolidated was maintained for two or three years before it was abandoned. The LB-XX designation stood for land bomber and the numbers were assigned sequentially.

The first single-engine attack aircraft was the LB-6 and its brochure was dated 22 September 1938. It had a gross weight of 15,000 pounds. The engine was buried in the fuselage and the engine type is unknown but probably was liquid-cooled. It appears to have had a crew of two or possibly three as evidenced by the bomb and gunnery stations. No other data is available. The second single-engine design, the LB-13, is noted in a studies index as a Scout Bomber with a gross weight of 10,000 pounds. The date of this design was 2 December 1938. It was of conventional design and also appears to have been powered with a liquid-cooled engine.

The first of the multi-engine attack bombers was the LB-8 dated 31 October 1938. This airplane was designed in response to the Army Air Corps Specification C-103. It had two Allison 1,000-hp V-1710-F1 liquid-cooled engines, a crew of three, and was configured to carry

The LB-14 Attack Bomber of January 1939 was designed to Air Corps Specification C-104 dated 13 September 1938. It is not known if this specification resulted in a design competition or a subsequent selection. The configuration was generally similar to LB-12 design, except it was somewhat larger and featured Allison liquid-cooled engines. The design incorporated the Consolidated trademark twin tail. Also studied was an LB-15 with the same airframe but with an alternate engine that was labeled a "flat" engine, indicating it was a horizontally opposed, or possibly the Pratt & Whitney X-1800 engine. (Convair via SDASM)

A second single-engine attack bomber was the LB-13 reported in December 1938. It was designated a Scout Bomber and had a gross weight of 10,000 pounds. The design of this airplane appears to have been significantly more sophisticated than the LB-6, as perceived by this illustration. As with the LB-6 no further information is available. (Convair via SDASM)

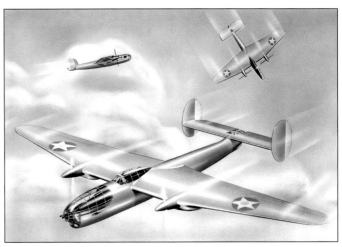

two 600-pound demolition bombs. Defensive armament included two fixed and four flexible guns, all .30 cal. It had a wing area of 450 square feet, a span of 60 feet, and a gross weight of 17,334 pounds. An alternate to the LB-8 was the LB-9, a three-engine version, again with the same Allison engines. Its wing area was increased to 527 square feet, its wingspan to 65 feet, and its gross weight to 22,350 pounds. The bomb load, crew, and defensive armament were the same as the LB-8. No drawing or image is available.

The LB-12 also seems to meet the same Air Corps C-103 requirements as the LB-8 and LB-9, but this version was slightly larger and used the Rolls-Royce Merlin 1,050-hp engines. It was also in the same time period, dated 1 November 1938. The LB-12 had a wing area of 550 square feet and a gross weight of 20,200 pounds. Other notes indicate a top speed of 352 mph and a range of 1,580 miles.

A new Air Corps Requirement, C-104 dated 13 September 1938, resulted in the LB-14 design of January 1939. This appears to have been a slightly larger version of the LB-12 but used the much larger Allison 2,000-hp V-3420-A3 powerplant. This design also considered alternate engines including the Pratt & Whitney X-1300 and the Wright A617. The gross weight of this design was 22,650 pounds and the wing had an area of 600 square feet and a span of 60 feet. It had the same crew, bomb load, and defensive armament as the previous attack bombers. A high speed of 430 mph was estimated for this Allison-powered design and it had a range of 1,200 miles. Added integral fuel increased the range to 1,960 miles. The high performance of the LB-14 was attributed to thin wings, low frontal area, Fowler flaps, and flush riveting. It was also estimated that projected engine performance improvements would provide an increase in the maximum velocity of 470 mph. The LB-15 is another version of this design with an alternate engine installation. The study index noted a "flat" engine that might have been the Pratt & Whitney X-1300. The gross weight of the LB-15 was 23,093 pounds.

The twin-engine LB-19 Attack Bomber was designed to a revised Air Corps Specification C-103A and this study was completed in March 1939. The engines that were considered for the LB-19 were the Pratt & Whitney R-2800 and the Wright R-2600, both in two-stage and two-speed alternatives. The LB-20 of April 1939 was a

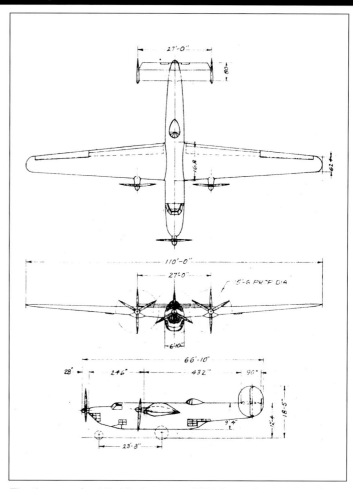

The LB-14 airplane had a gross weight of 22,650 pounds, and could carry up to 2,000 pounds of bombs. It had a top speed of 430 mph at 15,000 feet achieved by significant drag reduction via a low frontal area fuselage, flush riveting construction, and what is termed thin wings. It had a tricycle landing gear and used Fowler flaps that reduced the required wing area. Its range was 1,200 to 1,960 miles depending on bomb loading. (Convair via SDASM)

The three-engine LB-4 had a span of 110 feet, a length of 66 feet 10 inches long, and a gross weight of 50,000 pounds. The powerplants were the same as the twin-engine aircraft, the Allison XV-3420. Defensive armament included a turret on top of the fuselage, a tail turret, and a ventral gun position. No documentation as to conclusions of the relative merits of two versus three engines was ever found. (Convair via SDASM)

LB-19 and LB-20 represent another design cycle for an attack bomber conducted several months after the previous designs. The LB-19 shown here (March 1939) is a refined design of the same general configuration of their attack bombers. The aircraft had a gross weight of 21,083 pounds. A weight report indicates that several engines were considered including the R-2800, R-2600, and R-3350. (Convair via SDASM)

Bomber Programs

twin-boom version of the above noted Requirement for comparison with the more conventional LB-19 design. The LB-20 had the same baseline R-2800 engines and alternates and also had the same wing area. Gross weight was only slightly heavier at 21,147 pounds. No conclusions were noted in the available documentation with regard to the comparison of the twin-boom design and the conventional one.

The first medium-bomber study for which documentation survived is the LB-4, a design completed in August 1938. Both a two-engine and a three-engine version were investigated under the LB-4 designation. Both of these airplanes were designed for a gross weight of 50,000 pounds, a weight that somewhat later became the four-engine heavy bomber of the 1940s. The three-engine LB-5 design had the third engine installed just behind the cockpit and slightly ahead of the wing that drove the forward propeller with a shaft that ran through the cockpit. This differed from the previous three-engine designs that had the engine installed just in front of the cockpit. This LB-5 arrangement may have been aimed at shifting the engine weight aft to assist the aircraft's balance. The other two engines were buried in the wing. All engines in both versions of the LB-5 are believed to have been Allison XV-3420s.

Correspondence has recently been uncovered relative to the Air Corps' evaluation of the LB-5 three-engine proposal. Wright Field's Engineering Division report in September 1938 was scathing. The general arrangement was roundly criticized and the performance estimates were

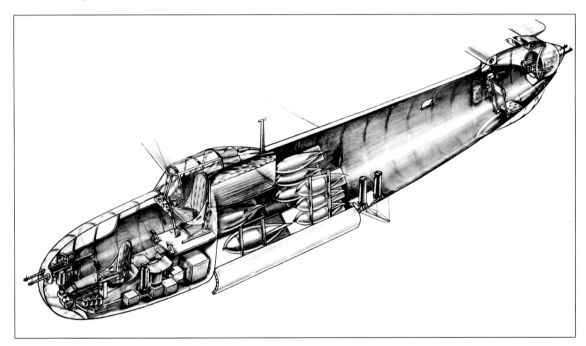

The LB-14 had a crew of three: bombardier/gunner in the nose, pilot, and tail gunner. This design had fixed twin guns in the nose position and a turret with two guns in the tail. The bomb bay was immediately behind the pilot. (Convair via SDASM)

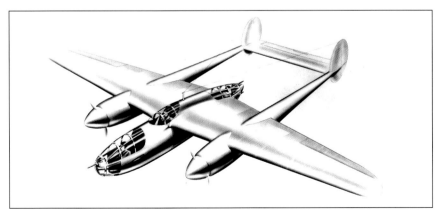

This twin-engine twin-boom attack bomber, the LB-20, was designed in April 1939, and is believed to have been for comparison with the previous LB-19. Engine installation was identical and the same engines were used. The LB-20's gross weight was 21,147 pounds, slightly more than that of the single-fuselage LB-19. It also had a crew of three and the same slim fuselage cross-section. Consolidated used this twin-boom-tail concept on several designs but it was never adopted. (Convair via SDASM)

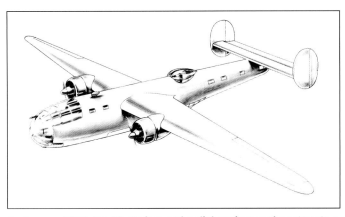

Engine installation may be seen in this cutaway drawing with the nose engine mounted in the fuselage behind the flight deck and driving the propeller by means of an extension shaft through the cockpit and between the two pilots. The motivation may have been to move engine weight aft. Allison V-3420 engines in the wing installation were completely buried. Forward and aft ventral fuselage gun positions were also visible and were flexible gun mounts only. (Convair via SDASM)

In January 1939, the LB-17 featured radial engines and a return to a circular fuselage cross-section, as well as the rounded nose, which looked very much like the B-29. There was a top fuselage gun turret, a tail turret, and a ventral mid-fuselage gun position. The aircraft's gross weight was 41,629 pounds, putting it in the heavy bomber class. It was apparently configured for comparison with the LB-16 four-engine bomber. (Convair via SDASM)

The LB-26 was apparently a smaller, lighter evolution of the LB-24 with a 579-square-foot wing, a span of 70 feet, and a length of 55 feet 3 inches. Gross weight was 22,500 pounds, about 5,000 pounds less than the LB-24. The LB-26, or possibly the LB-24, may have been Consolidated's proposal for the Air Corps' medium bomber that was awarded to North American's B-25 and Martin's B-26, as they were in the same weight class and time period. (Convair via SDASM)

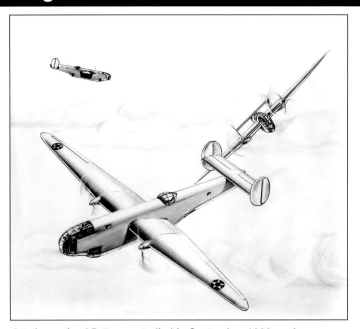

A twin-engine LB-5 was studied in September 1938, and was a smaller medium bomber of 35,000 pounds gross weight. It had a narrow, deep fuselage and two buried engines in the wing, probably the same Allison engines that were then favored by Consolidated. The fuselage had a rounded nose and the same fore and aft ventral gun positions as the LB-4 three-engine design. (Convair via SDASM)

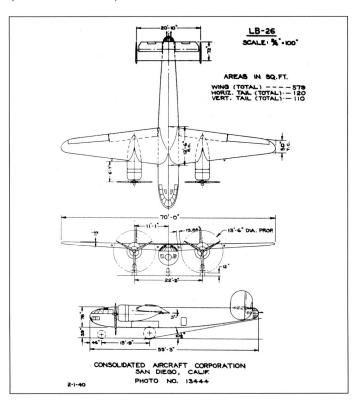

Bomber Programs

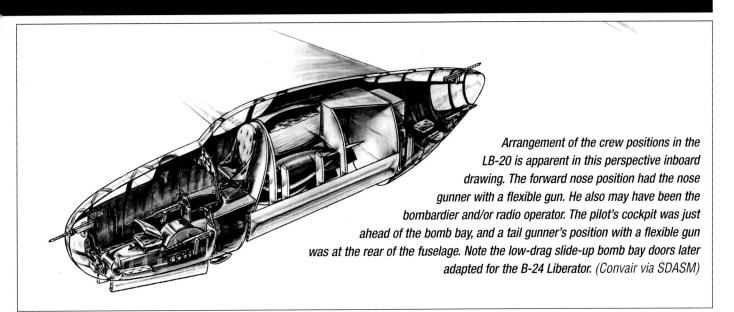

Arrangement of the crew positions in the LB-20 is apparent in this perspective inboard drawing. The forward nose position had the nose gunner with a flexible gun. He also may have been the bombardier and/or radio operator. The pilot's cockpit was just ahead of the bomb bay, and a tail gunner's position with a flexible gun was at the rear of the fuselage. Note the low-drag slide-up bomb bay doors later adapted for the B-24 Liberator. (Convair via SDASM)

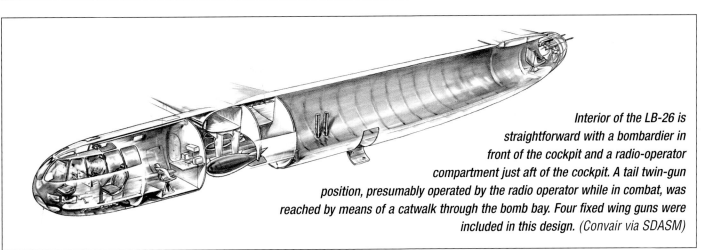

Interior of the LB-26 is straightforward with a bombardier in front of the cockpit and a radio-operator compartment just aft of the cockpit. A tail twin-gun position, presumably operated by the radio operator while in combat, was reached by means of a catwalk through the bomb bay. Four fixed wing guns were included in this design. (Convair via SDASM)

noted as exceedingly optimistic. They did, however, recommend continuing the LB-5, but confined to a two-engine version. This was apparently Consolidated's motivation for undertaking work on that configuration.

The twin-engine version was also designed with the buried wing engines and a rounded nose. The buried engines and the circular fuselage lent themselves to a very clean aerodynamic design.

The LB-5 study, completed a month later in September, was a twin-engine design that was somewhat lighter, with a gross weight of 35,000 pounds. This configuration also had buried liquid-cooled engines in the wing, a rounded nose, and the Consolidated twin tail. It also had a deep, narrow fuselage with flexible gun positions in both the fore and aft ventral positions and

probably a tail turret. No other information regarding this design study or its motivation is available.

The next twin-engine medium bomber, the LB-17 baseline design, dated 20 January 1939, used the Wright R-3350-A614 turbo engines. This configuration was to be compared with the four-engine LB-16. Alternate engines investigated for this two engine study included the R-3350 in both the single stage and the two-speed versions, but it was noted that these were not submitted to the customer. The payload of the LB-17 was the standard four 600-pound bombs. It had a crew of nine, and the defensive armament included three .50-cal. and two .30-cal. guns, all with flexible mountings. The gross weight of this aircraft was 41,629 pounds, again the weight class of heavy bombers of World War II.

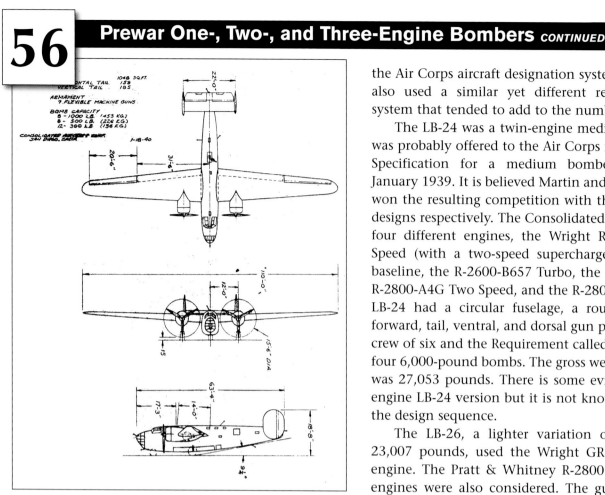

The LB-29 proposed the conversion of the B-24 airframe to a twin-engine configuration using the larger engines. Similar proposals for the B-24, PB4Y, and B-32 were thwarted by the demand for the big engines by other programs. The LB-29 had the basic XB-24 dimensions: length of 63 feet 4 inches and 1,048-square-foot wing with a span of 110 feet . Maximum bomb load capacity of the LB-29 was 8,000 pounds, as was the B-24. (Convair via SDASM)

The LB-22 was a two-engine bombardment aircraft configured in response to Air Corps Specification C-213, and presented in the aircraft brochure that was dated 7 May 1939. The specifics of this requirement are not known, but it followed the B-24 specification that was C-212. This airplane was a derivation of the basic and final XB-24 design whose contract had been negotiated two months earlier.

The next twin-engine medium bomber was the LB-24 initially offered in a brochure dated 8 February 1939 and revised in June 1939. The LB-24 should not be confused with the B-24 where Consolidated was using a predesign designation system that was somewhat in conflict with

the Air Corps aircraft designation system. Consolidated also used a similar yet different report numbering system that tended to add to the number confusion.

The LB-24 was a twin-engine medium bomber that was probably offered to the Air Corps in response to its Specification for a medium bomber published in January 1939. It is believed Martin and North American won the resulting competition with the B-26 and B-25 designs respectively. The Consolidated design looked at four different engines, the Wright R-2600-B655 Two Speed (with a two-speed supercharger) that was the baseline, the R-2600-B657 Turbo, the Pratt & Whitney R-2800-A4G Two Speed, and the R-2800-AG Turbo. The LB-24 had a circular fuselage, a rounded nose, and forward, tail, ventral, and dorsal gun positions. It had a crew of six and the Requirement called for a payload of four 6,000-pound bombs. The gross weight of the LB-24 was 27,053 pounds. There is some evidence of a four-engine LB-24 version but it is not known how it fits in the design sequence.

The LB-26, a lighter variation of the LB-24, at 23,007 pounds, used the Wright GR-2600 two-speed engine. The Pratt & Whitney R-2800 and the R-1830 engines were also considered. The guns were revised with four fixed guns added in the wing, and an optional tail turret or alternately, top and bottom fuselage gun positions, were offered. The crew was reduced to four. The LB-26 had a top speed of 326 mph

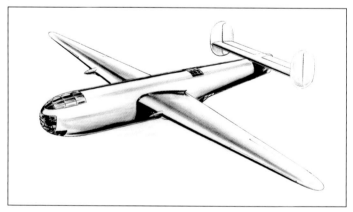

Earliest of the medium-bomber designs dated August 1938 was the LB-4 where both a two-engine and a three-engine version were considered. The twin-engine LB-4's mid wing had a very clean buried Allison XV-3420 liquid-cooled engine installation and had the Consolidated twin tail also. The aircraft had a gross weight of 50,000 pounds and a top speed of 320 mph. (Convair via SDASM)

Bomber Programs

The LB-28, designed in February 1940, was an alternate version of the LB-27, possibly an export version that used the larger Wright R-3350 engines with an interesting engine nacelle and propeller design. Canopy and nose have been revised, fuselage lengthened, and fuel capacity added. Note gun position aft of the cockpit. Gross weight was 24,443 pounds, and the increased fuel capacity and larger engines allowed doubling the bomb load and a significant increase in range. (Convair via SDASM)

and a range of 1,000 miles. It is not known how this version related to the LB-24 or if this was the final proposal configuration or an alternate.

An export version of the LB-19 was offered as the LB-27 in a brochure dated 9 February 1940. It was stated that the equipment was similar to the LB-26, had a crew of four and a bomb payload of 1,200 pounds. The engines offered were the Wright R-2600, the Pratt & Whitney R-2800, and the P&W R-1830. Armament included four fixed guns, two in the wing and two in the fuselage, and three flexible guns in the fuselage, all .30-cal. The gross weight of the R-2600-powered airplane was 20,759 pounds.

The LB-28 was basically the same as the LB-27 export bomber but had a longer fuselage, revised nose and canopy, and offered a different powerplant, the

Wright R-3350 two-speed. It carried increased fuel providing for increased range, and had double the bomb load capability. It also had a crew of four and the same gun arrangement as the LB-27.

The LB-29, offered 7 March 1940, was a twin-engine export version of the XB-24, as the latter program was just getting underway. The engines in this variation were the 2,000-hp Wright R-3350 Two Speeds in lieu of the four Pratt & Whitney R-1830s. The LB-29 was identical to the XB-24 except for those changes necessary for the different powerplant. It was configured for a crew of six, had three flexible .50-cal. guns, and four .30-cal. guns. It carried four 600-pound bombs, had a gross weight of 35,353 pounds, and a range of 2,960 sm at its most economical cruising speed. The top speed of the LB-29 was 361 mph.

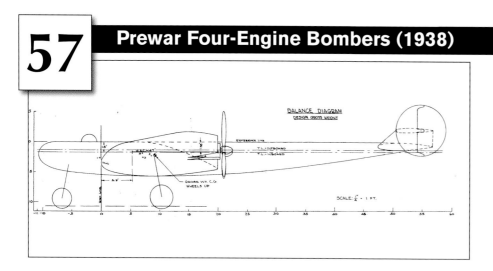

BALANCE DIAGRAM
DESIGN GROSS WEIGHT

SCALE: ¼" = 1 FT.

Two four-engine bombers were studied in March 1938. Essentially the same design, one was a pusher configuration and the other had tractor engines. These two designs represented an evolution away from the flying-wing configuration and now incorporated a full conventional fuselage. The report that included this rudimentary drawing indicated the study was in response to the Army Air Corps Specification 98-208. (Convair via SDASM)

Consolidated had considered the radical-design flying wing during 1937 and 1938, resulting in numerous two- and four-engine configurations, both for military bombers and civilian applications. The flying-wing design underwent continual evolution toward more and more conventional versions and Consolidated abandoned the concept by early 1938.

The first of the subsequent four-engine airplanes studied, of a much more conventional design as revealed in the archives, was the LB-380014 and LB-380015. These designations might relate to a drawing number system, indicating the 14th and 15th designs or drawings of 1938. In any case these four-engine bombers show the clear evolution away from the flying wing and toward a full-fuselage aircraft. The LB-380014 used a pusher-engine arrangement and the LB-380015 was a tractor version of the same design. Only a weight report with a profile drawing survived and no other information is available.

The next design was the LB-16 in January 1939. It was a straightforward shoulder-wing configuration with Consolidated's trademark twin tail. This design retained the rounded-glass nose that had been included on many other designs of the period. It included a large turret on top of the aft fuselage and a nose, tail, and (apparently) ventral gun position. It appears to have been compared with the LB-17, a twin-engine design that undoubtedly had the larger engines, as both had about the same gross weight. No conclusions from this comparison are available.

The four-engine designs being considered had evolved somewhat by February 1939 and show a revised rounded nose, a circular fuselage cross-section, and revised gun stations. The most significant aspect of this design is that it was, for the first time, designated the Model 32, which was soon to become the B-24.

By May 1939, the rounded nose was still retained in the design, as were nose and tail guns, but the fuselage waist gun positions were no longer included. It also appears that the fuselage design was starting to acquire the deeper fuselage cross section later associated with the B-24, rather than circular design. At this time a full-scale mockup of the Model 32 in this configuration was well underway indicating that Consolidated was becoming confident that there would be hardware forthcoming.

Only one month later, in June 1939, documentation showed the final familiar design of the B-24 with the deep fuselage and the stepped cockpit arrangement. It was referred to in brochures (that, unfortunately, are no longer available) as, in one case, the LB-24 with four engines, and in another case, the 32-X and the XB-24. It is believed the LB-24 four-engine designation may have been a misnomer, as the LB-24 design that was found was a twin-engine medium bomber with a gross weight of 27,000 pounds. The brochure titled 32-X also included in its title a reference to the Air Corps Specification C-212 that defined the planned heavy-bomber procurement.

A further uncertainty, complicated by fragmentary documentation, is the LB-22, a twin-engine version of the Model 32. This design used the Allison V-3420 liquid-cooled engines and the aircraft had essentially the same gross weight as the Model 32, clearly putting it in the heavy-bomber category. The brochure on this design is not in the archives but the title from an index includes reference to Air Corps Requirement Specification C-213. It is not known what this specification covered, following as it did, the C-212 for the B-17/B-24 heavy-bomber program.

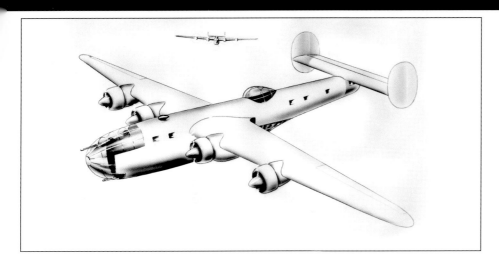

Consolidated's next four-engine bomber study, after the flying wings, was the LB-16, published on 20 January 1939. It was a fairly conventional configuration and had a rounded nose, as was the case with many designs of the period. It had a crew of nine, was designed for a bomb load of four 600-pound bombs, and had a gross weight of 41,000 pounds. Defensive armament included three .50-cal. and two .30-cal. flexible gun positions. LB-17 was a similar comparison design but with twin engines. *(Convair via SDASM)*

By 16 February the design had evolved somewhat but still retained the rounded nose. By now the heavy-bomber program was starting to take shape and the design had been designated Model 32. Fuselage was circular and had shed the large turret-top enclosure and the ventral gun position of the LB-16. The engines are believed to have been the Pratt & Whitney R-1830, common to these four-engine designs. A full-scale mockup of the Model 32 was under construction at the time. *(Convair via SDASM)*

This full-scale mockup of a round-nose Model 32 is believed to have been completed by May or June 1939 and is shown here in the Consolidated plant. The configuration appears to be intermediate to the previous two designs. The deeper fuselage had the bottom contour of the LB-16 and the large upper-fuselage gun position but had the engine nacelles and possibly the Davis wing of the February design. In any case the design details gave way to the final XB-24 configuration and fabrication was begun. *(Convair via SDASM)*

XB-24 is shown in flight over the San Diego area in late January 1940, about one month after its first flight on 29 December 1939. Note the lack of markings including the tail number, except for the rudder, which had been painted. Only a few design changes are apparent from the previous artist rendition depicting the configuration prior to its manufacture. Minor differences appear mainly in the engine nacelle area. (Convair via SDASM)

Bomber Programs

The B-24 story began with the Army Air Corps program for a heavy bomber centered on the B-17. The XB-17 had crashed in October 1935 but the Air Corps, eager to get this program underway, contracted for 17 YB-17s that were to be delivered by the fall of 1938. With the inevitable war looming closer, the Air Corps was deep in buildup planning and was worried about production capacity. Gen. Hap Arnold contacted Maj. Fleet, an old friend, and suggested that Consolidated also build the B-17 under license to Boeing. Consolidated's I. M. "Mac" Laddon and C. A. Van Dusen did indeed visit Boeing but Fleet didn't think the company should build a four-year-old design.

The Air Corps had just issued Specification C-212 in December 1938 for a bomber capable of 3,000 miles range with 8,000 pounds of bombs, which could achieve 300 mph and 35,000 feet. Consolidated had been studying similar requirements for several years and felt that by using the more efficient Davis wing the design could significantly improve the range with the same payload. Consolidated took its preliminary data back to Wright Field at the end of January 1939. The remaining competitors for the C-212 procurement, Martin and Sikorsky, didn't have designs that were sufficiently mature. Consolidated and Boeing were therefore the winners of this heavy-bomber procurement.

The B-24 configuration had evolved from the LB-16 and subsequent studies of 1939 and 1940. Consolidated was quite confident that it would receive a contract and it built a full-scale mockup of the rounded nose airplane based on the Model 32 design as it existed in early 1939.

The B-24 design, with its distinctive Davis wing and the twin tail, had many innovations. These included the deep fuselage that facilitated bomb storage and handling, roll-up bomb bay doors, tricycle landing gear, Fowler flaps, and integral fuel tanks. The wing area was 1,048 square feet with a span of 110 feet and a length of 66 feet 4 inches. The story goes that Maj. Fleet, in his inimitable fashion, commented at the XB-24's rollout that 3 feet should be added to the nose, "so it would look better," and it was done. The gross weight of the XB-24 was 41,000 pounds and four Pratt & Whitney R-1830 engines powered it. The size and weight of the designs for these heavy bombers remained consistent over all of the variations of the several years of study.

By February 1939 the mockup was well along and on 30 March Consolidated received a contract for seven YB-24 (Model 32) airplanes for a total of $2.88 million with a first flight to take place in nine months. In April, Congress approved the Air Corps' expansion to 6,000 airplanes and Consolidated was then negotiating an additional quantity of 38 B-24s. On 10 August Consolidated received orders for the B-24As previously negotiated along with Boeing, which received a contract for 38 B-17Cs. The XB-24 flew for the first time on 29 December 1939, nine months from contract initiation.

The French had ordered 175 LB-30s, the export version of the Model 32, but after their collapse the order was taken over by the British. Because of the urgent need, the Air Corps permitted them to exchange some of the later scheduled LB-30s for earlier available YB-24 and B-24A models. The British accepted their first airplane in January 1941. LB-30 was the last predesign designation in the land bomber series.

With an enormous war production buildup underway, three additional companies were brought in to also produce the B-24. They were Ford in Willow Run, North American in Dallas, and Douglas in Tulsa. All told there were 18,481 B-24s built in the almost five years of production, more than any other American airplane. Production in Consolidated Vultee's San Diego plant numbered 6,724 aircraft.

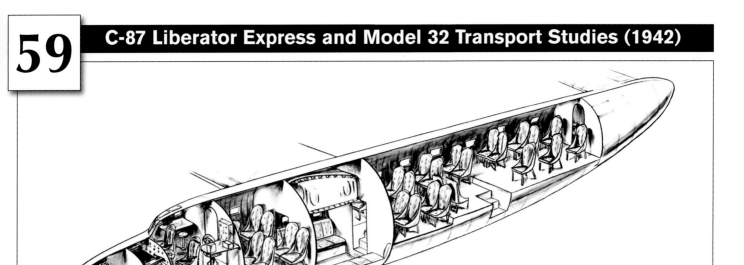

Consolidated was looking at transport versions of the Model 32 as early as the 1940s. This troop transport concept of October 1941 incorporates a new optimized fuselage to avoid the limitations of the bomber design. This configuration carried 40 troops for a maximum range of 3,300 miles at an average cruising speed of 185 mph. Although it is stated that this concept would accommodate 40 troops, the seating in this illustration appears to be for a maximum of 38, depending on the number in the center section. It also illustrates the "Achilles' heel" of the Model 32 design in that the wing carry-through structure is very intrusive for a transport fuselage. A version of this concept with a slightly longer fuselage was investigated for commercial applications. (Convair via SDASM)

The first studies of a commercial version of the Model 32 appear to have taken place in early 1940 and military conversions were undoubtedly also studied in this time period. The first available documentation of a Model 32 military transport study appeared in mid October 1941, about 10 months after the first flight of the XB-24. This adaptation of the bomber design had an all-new fuselage with a circular cross section probably in early recognition of the limitations of the bomber's fuselage design. It was termed a troop transport and was claimed to accommodate 40 passengers and a crew of four, with a range of 3,300 miles and cruising speed of 185 mph.

The first proposals and variants of transport conversions of the B-24 occurred early in the B-24 production program prior to the entrance of the United States in World War II. The first modifications offered included a minimal "quick change" conversion and a "regular" conversion with more extensive changes. Several of the British LB-30s from the early Model 32 production were converted with the quick change. It is also believed a small number of early B-24s were modified with a regular conversion for cargo and passenger service prior to the advent of production of the C-87 and the C-109.

The prototype C-87 was an LB-30B that had sustained damage and was rebuilt in San Diego as an XC-87. It first flew on 24 August 1942 and the first production C-87, a 20-passenger version, was delivered in September 1942 at Fort Worth. This aircraft was, in general, similar to the "regular" change configuration but was newly manufactured and was not a modification. Convair built 286 C-87s mostly at the Fort Worth plant on the B-24 line.

The C-109, a Ford-modified B-24E, was a C-87 type to be used in the China-Burma-India (CBI) Theater that is believed to have retained the tail turret and had provisions for extra fuel. Ford converted 199 B-24s and Martin converted 9 B-24s to the C-109 configuration. The Navy also received three VIP conversions of the C-87 as the RY-1, and six C-87s as RY-2s.

A C-87B was offered in September 1943 and a C-87C was offered in March 1944. The latter was supposedly an Air Force version of the Navy's RY-3. The only information available on the C-87B was an illustration that appears almost identical to the C-87C. Neither version was procured.

Bomber Programs

First B-24 transport, the XC-87, was converted in San Diego from an LB-30 that had been involved in an accident. The XC-87's first flight occurred on 25 August 1942. All subsequent C-87s were built on the B-24 production line at the Fort Worth plant with a total of 286 produced including this unpainted example. (Convair via SDASM)

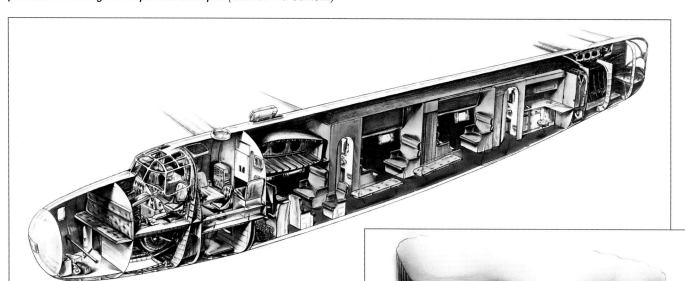

Some of the early C-87s were converted for VIP passenger transport use both for the Army and the Navy as the RY-1. These drawings, from November 1942, may illustrate such an application. This interior had two private compartments, a private restroom, a washroom toilet, and a galley. It appears that it would accommodate eight passengers not including those who might use the folding bench seats. Interestingly, there are provisions for a singe .50-cal. tail gun included in this concept. (Convair via SDASM)

Private sleeping compartment in a night conversion arrangement included a private restroom and a coat closet for the VIP. It is believed that the configurations illustrated may have been intended for either the military or for ranking government personnel as indicated by the civilian clothes in the coat rack. (Convair via SDASM)

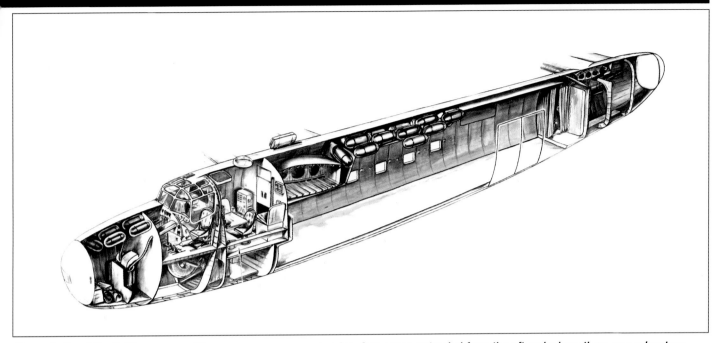

The majority of the C-87s were used for cargo or passenger service. Cargo area extended from the aft end where there was a lavatory, forward to the bulkhead at the flight deck just in front of the wing structure. A loading and entrance door was located on the port side. Personnel capacities of from 18 to 26 have been noted, with 20 passengers being the most prevalent. *(Convair via SDASM)*

C-87C TRANSPORT

CONSOLIDATED VULTEE AIRCRAFT CORPORATION
DEVELOPMENT ENGINEERING   SAN DIEGO, CALIF.

PHOTO NO. 1410B   DRG. NO.101 Z 003
REF. 101 Z 001

7P 101 002   3-1-44

A C-87C was offered in 1944 but was not built. It was reportedly an Army version of the RY-3 that was in turn a transport version of the PB4Y-2. It was visually similar to a C-87B previously proposed in September 1943, also not built, except for a more square "Convair" single vertical tail in lieu of the B-29 vertical fin considered earlier. This aircraft was believed to have had a capacity of 30 passengers. *(Convair via SDASM)*

A brochure dated March 1941 summarized a commercial passenger version of the Model 32. It had seating for 18 passengers in individual reclining seats or a total of 22 if the lounge seating of four was eliminated. It had a range of 1,650 statute miles at a cruising speed of 221 mph and a top speed of 295 mph. Gross weight was 45,310 pounds. This illustration has markings very similar to those of American Airlines at the time. (Convair via SDASM)

A commercial passenger version of the Model 32 was documented in a brochure published 25 March 1941. Some of the data was dated March 1940, indicating a year of activity prior to the brochure. This design (LC-7) was a direct conversion of the Model 32 airframe with the identical fuselage and would accommodate a crew of five, 18 passengers in reclining seats, and four passengers in a lounge.

About a year after the Model 32 military transport that featured the new circular-fuselage conversion was studied, a commercial version was also examined in December 1942. The design had evolved somewhat in that the fuselage length had been increased by 13 feet to almost 81 feet and the commercial passenger capacity was reduced to 33.

The attempts at commercialization of the B-24 were, of course, put on hold by the war, and little activity occurred until 1944 when the company's venture, the Model 39, first flew on 15 April. That airplane incorporated a new circular fuselage mated with the PB4Y-2 wing, tail, and landing gear, and would accommodate 48 day passengers. The design generated little interest and the Navy canceled its order for a quantity of cargo versions. Unfortunately the Model 32 design and its variants had a serious design disadvantage, discouraging its use for transport applications. As a high-wing design, the wing carry-through structure effectively divided the fuselage into two sections that were not reasonably accessible to each other.

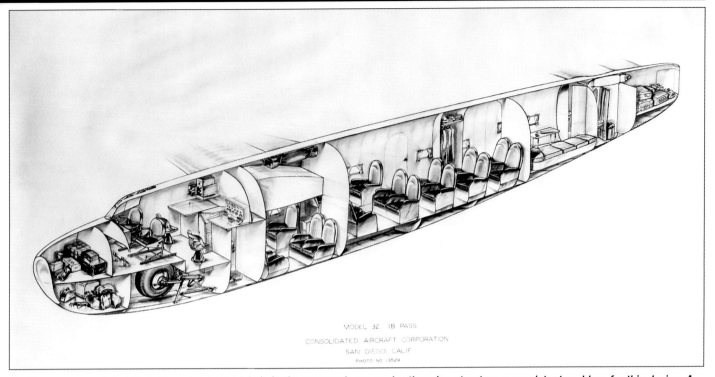

MODEL 32 18 PASS
CONSOLIDATED AIRCRAFT CORPORATION
SAN DIEGO, CALIF
PHOTO NO. 13529

Seating accommodated 12 in the main area and six in the cramped area under the wing structure, a persistent problem for this design. A lounge that seated four was located aft of the main area. Restroom was in the aft fuselage and baggage was carried in compartments in the nose and far aft of the fuselage. (Convair via SDASM)

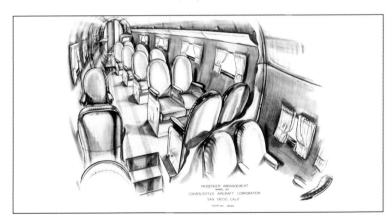

PASSENGER ARRANGEMENT
MODEL 32
CONSOLIDATED AIRCRAFT CORPORATION
SAN DIEGO, CALIF.
PHOTO NO. 13542

This view into the main cabin illustrates that at the maximum width the fuselage could accommodate only a two-and-one seating capability, which was quite limiting. The seats in the foreground were actually under the wing structure. A lounge and a restroom were at the rear. (Convair via SDASM)

In this design, the wings, powerplants, landing gear, and empennage were all directly taken from the Model 32. The new 10-foot 6-inch circular fuselage was lengthened by 13 feet from that of the troop transport studied earlier in October 1941 that also had the new circular fuselage. Gross weight of this airplane was 56,000 pounds, a significant increase from the Model 32 at the time. (Convair via SDASM)

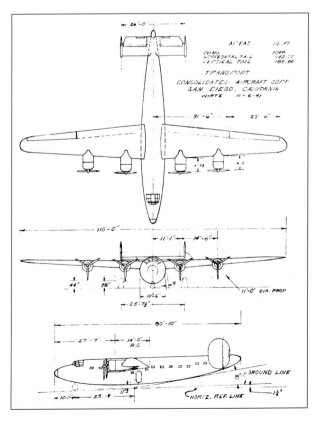

CONSOLIDATED AIRCRAFT CORPORATION
SAN DIEGO, CALIF.
PHOTO NO. 13503

The LB-25 design, dated 6 May 1940, is believed to have been the going-in configuration for the R-40B competition. It was heavily armed, including retractable top and bottom turrets, and a four-gun tail turret. In addition there were two remote-controlled twin gun turrets housed in the aft end of the inboard engine nacelle and also two remote-controlled forward-firing guns located in the wing leading edge outboard of the engines.
(Convair via SDASM)

The origins of the Model 33 date back to the early 1930s when the Army Air Corps was establishing a general requirement for a very long-range bomber with a range of 5,000 miles. This program initially yielded the two experimental aircraft, the Boeing XB-15 that was first flown in August 1937 and the Douglas XB-19 flown in June 1941. Both of these efforts were beyond the supporting technology, particularly in regard to engines. In its next effort the AAC initiated requests for information in October 1937 for a new "Super Bomber" to replace the B-17 that would be bigger, have increased performance, be pressurized, and have supercharged engines. Industry response was not enthusiastic in that all government contracts at that time were fixed-price and, lacking advance payment, the companies had to fund their own work. Four companies, reportedly including Consolidated, responded with rather ordinary concepts. Consolidated's submission has not been identified but may have been the LB-380014 and LB-380015 studies that were conducted during the same time period.

As the Air Corps' B-17 and B-24 production programs were getting underway, this Super Bomber program was more or less dormant until late 1938. In June 1939 the AAC recommended a program that included a bomber with a 5,500-mile range and new engines to make this performance possible. Gen. "Hap" Arnold, the recently appointed head of the AAC, was able to secure funding in November 1939, and the R-40B requirements were issued dated 8 April 1940. The R-40B, together with Specification XC-218, included a range requirement of 5,333 miles, a 2,000-pound load for the full range, self-sealing fuel tanks, increased armor, and multiple defensive gun turrets. Four companies, Lockheed, Douglas, Consolidated, and Boeing, responded immediately and Martin later joined the list. All were awarded small contracts for mockups and data. Lockheed and Douglas essentially dropped out and Martin was given a separate contract.

Consolidated's LB-25 study concept was its initial configuration for a heavy bomber in the 80,000- to 90,000-pound class using the Wright R-3350 engines and was to be significantly advanced from the then-current

Bomber Programs

B-17 and B-24. The first study reported in the records is not available but was dated 5 March 1940, shortly before the R-40B requirements were issued. The design at that time had a gross weight of 85,600 pounds. Several drawings and an artist rendition from the same time period reveal the main characteristics of the configuration at that point in its evolution. The earliest drawing available is of a wind tunnel model from October 1939, and shows a four-engine, high-wing, twin-tail, circular-fuselage design, visually not too different than the B-24. The nose had a distinct B-24 look and the twin vertical stabilizer was distinctly Consolidated. A drawing in late February 1940, four months later, shows a lengthened fuselage and a revised, more rakish, vertical-fin shape. It is this configuration shown in an illustration that appears to have been the R-40B proposal concept submitted on 8 April 1940.

By February 1940, the LB-25's wingspan had been increased by 10 feet to 135 feet, providing a total area of 1,422 square feet and was the final wing design. Vertical tail surfaces had acquired an uncharacteristic rakish back sweep that lasted until the XB-32 configuration was finalized and the tail reverted to the traditional Consolidated design. Fuselage at this point also featured a smaller clear nose section. (Convair via SDASM)

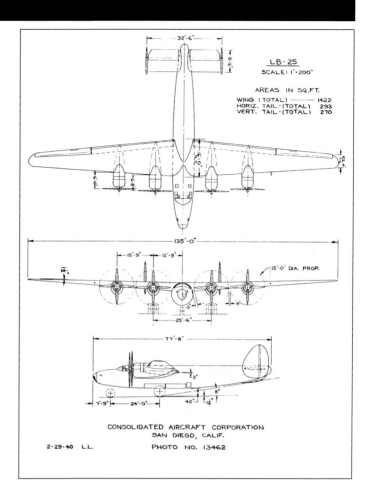

62 | B-32 Dominator - Model 33 (1940)

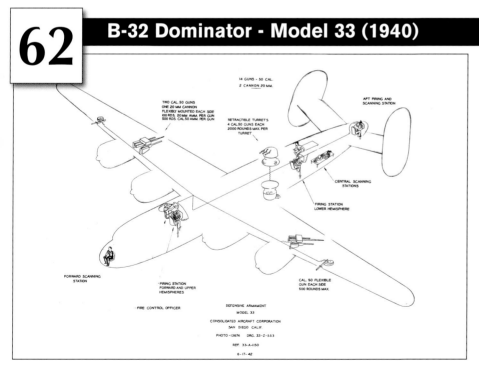

This graphic illustrates the initial defensive armament system proposed for the B-32 bomber in response to the R-40B requirements. Gunners were located at scanning stations in the nose, at a forward upper firing station, at a lower station in the rear fuselage, and in the tail. Guns included an upper and ventral retractable turret with four .50-cal. guns each and two turrets in the rear of the outboard engine nacelles with two .50-cal. and one 20mm gun in each. In addition there were two forward-firing flexible guns in the wing leading edge outboard of the outer engines. (Convair via SDASM)

Bomber Programs

XB 32
(CONSOLIDATED MODEL 33)

CONSOLIDATED AIRCRAFT CORPORATION
SAN DIEGO, CALIF.

The proposal configuration, designated XB-32 (Model 33) at the time of contract award, was developed from the LB-25 study concept. It had a rather streamlined circular cross section fuselage and defensive armament quite similar to the study concept. Small, clear nose and clear cockpit canopy, although they did not prevail in the B-32 design, foretold the trend that was to take place in the near future. (Convair via SDASM)

The two winners of the R-40B competition were Boeing, which received a contract for two XB-29 prototypes on 24 August 1940, and Consolidated, which received a contract for two XB-32s on 6 September. A third prototype was added to both contracts in December, and in June 1941 Consolidated was awarded a further contract for 13 YB-32s.

Consolidated provided a wind tunnel model with its proposal for test at Wright Field and based on these tests the configuration evolved somewhat away from the LB-25 design and toward what was becoming the XB-32 configuration. Although the Air Corps was concerned with the empennage design in terms of directional stability, as revealed in the wind tunnel tests, the twin tails were retained for the time being. Other various design modifications were made including a blunter nose, a more cylindrical fuselage, and Consolidated's less rakish tail. The revised mockup was completed and approved in December 1941.

The first XB-32 was completed on 1 September 1942 and made its first flight on 7 September. This first flight was a near disaster as very intense tail flutter was encountered, but the prototype was able to land safely. Many and continuing problems plagued the program

RECONNAISSANCE LANDPLANE
4-ENG.

MODEL 33

CONSOLIDATED VULTEE AIRCRAFT CORPORATION
DEVELOPMENT ENGINEERING   SAN DIEGO, CALIF.

PHOTO NO. 14793   DRG. NO. SD 45 10 300

In April 1945, Consolidated offered a reconnaissance landplane version of the Model 33 to the Navy in the manner of the PB4Y derived from the Model 32 (B-24). This version included three fuselage turrets, nose and tail turrets, as well as patrol bomber weapon systems such as radar and an anti-submarine warfare (ASW) suite. (Convair via SDASM)

limited combat in the Pacific, but the ending of hostilities ultimately led to the cancellation of the program on 18 September 1945 and complete termination on 12 October. Only three prototypes and 141 production aircraft had been built at contract termination, including 12 flyable but incomplete aircraft. The B-32 had initially been named the Terminator in late 1943 at Consolidated's suggestion, but the Air Corps' Aircraft Naming Board changed it to Dominator. This was later reversed by the State Department in the name of "political correctness" and Terminator was reestablished, although the name Dominator tends to persist.

Several variants and proposals of the basic B-32 include a paratroop transport conversion. It is also thought that the next B-32 production version would have a lengthened and revised nose section in response to comments from combat experience.

A major concern in the European Theater was the inability to maintain fighter escort for the full range of the strategic bombers. The concept was developed for a very heavily armed bomber whose mission was that of an escort aircraft to provide additional defensive armament. This led to the Consolidated XB-41, a version of the B-24 that could provide this capability. It was decided by the Air Corps that a longer-range fighter escort was the solution to this problem. An XB-32 escort was, however, proposed in March 1943 that had six powered turrets in the manner of the XB-41.

In October 1944, a turboprop-powered version of the B-32 was studied. Turbine propulsion was progressing and one of the first turboprops, the GE TG-100 (XT31), was in development. The resulting B-32 using these turboprop engines did not have significantly

leading to the cancellation of the YB-32 order. Consolidated's difficulties were compounded by the crash during takeoff of the first prototype on 10 May 1943 due to an apparent flap problem. This mishap in turn caused significant test program delays and seriously impacted the production program. Meanwhile the second XB-32 was completed in July 1943 and it retained the twin-tail design although directional stability was still a major concern. An attempt at resolving this problem included fitting a production B-29 vertical stabilizer to the XB-32 but directional and longitudinal problems persisted. The vertical-tail problem was resolved when the third XB-32, in mid 1944, was fitted and tested with a new, taller Consolidated vertical-fin design.

Because of the interminable delays in the B-32 program and the fact that the B-29 testing had demonstrated the ability to meet the Air Corps' strategic high-altitude Very Heavy Bomber requirements and was well into production, the B-32 was not needed as a backup and was downgraded in Air Corps planning. In essence it was relegated to use in the Pacific Theater as an advanced replacement for the B-24 for tactical-type missions. This again necessitated major redesigns and production delays but because of the major investment in the program it was not canceled. The B-32 did see

The second XB-32 is shown in flight over the San Diego area in July 1943. The first flight of the XB-32 occurred on 7 September 1942 and immediately encountered problems with the rudder and vertical fins. Other problems continued to plague the XB-32 test program culminating in the crash of the first prototype on 10 May 1943 with the loss of four lives. This led to significant redesign and the incorporation of a single, large vertical fin. *(Convair via SDASM)*

Production B-32s are recognizable by the redesigned nose and cockpit canopy and by the single tail. After the early difficulties with the B-32 twin-tail design and abortive testing with the B-29 tail, Consolidated designed an elongated single tail that solved the problem. *(Convair via SDASM)*

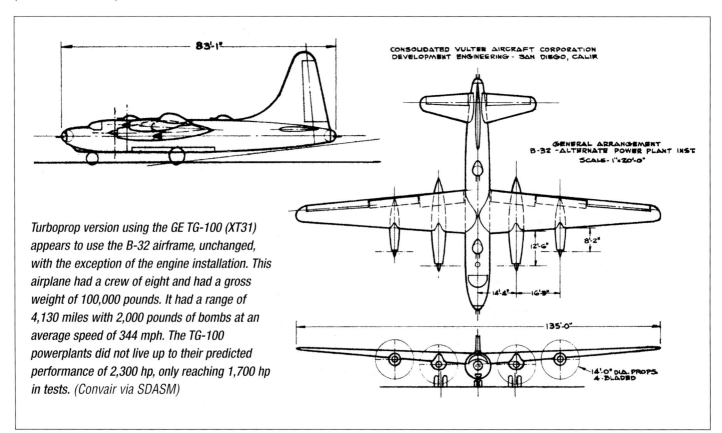

Turboprop version using the GE TG-100 (XT31) appears to use the B-32 airframe, unchanged, with the exception of the engine installation. This airplane had a crew of eight and had a gross weight of 100,000 pounds. It had a range of 4,130 miles with 2,000 pounds of bombs at an average speed of 344 mph. The TG-100 powerplants did not live up to their predicted performance of 2,300 hp, only reaching 1,700 hp in tests. *(Convair via SDASM)*

Bomber Programs

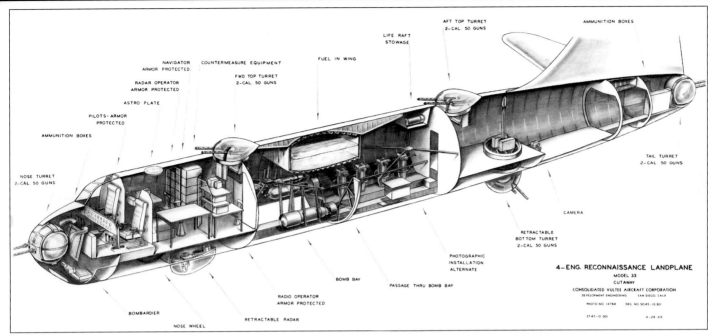

AFT TOP TURRET
2-CAL .50 GUNS

AMMUNITION BOXES

LIFE RAFT
STOWAGE

NAVIGATOR   COUNTERMEASURE EQUIPMENT   FUEL IN WING
ARMOR PROTECTED

RADAR OPERATOR   FWD TOP TURRET
ARMOR PROTECTED   2-CAL .50 GUNS

ASTRO PLATE

PILOTS- ARMOR
PROTECTED

AMMUNITION BOXES

TAIL TURRET
2-CAL .50 GUNS

NOSE TURRET
2-CAL .50 GUNS

CAMERA

RETRACTABLE
BOTTOM TURRET
2-CAL .50 GUNS

PHOTOGRAPHIC
INSTALLATION
ALTERNATE

4-ENG. RECONNAISSANCE LANDPLANE
MODEL 33
CUTAWAY
CONSOLIDATED VULTEE AIRCRAFT CORPORATION
DEVELOPMENT ENGINEERING   SAN DIEGO. CALIF

PHOTO NO. 14784   DWG NO. SD45-10.301

ZF45-10.001   4-28-45

BOMB BAY

PASSAGE THRU BOMB BAY

RADIO OPERATOR
ARMOR PROTECTED

BOMBARDIER

RETRACTABLE RADAR

NOSE WHEEL

Inboard profile of the Navy recon version shows the defensive armament with twin .50-cal. guns in nose and tail turrets, two twin .50 cal. guns in the forward and aft top turret, and a retractable bottom fuselage ball turret. A retractable radar unit was also housed in the fuselage bottom just aft of the flight deck. The large bomb bay accommodated photoflash bombs, depth charges, or mines. This particular design also shows the alternate installation of K-17 cameras. (Convair via SDASM)

This illustration of a twin-engine version of the Model 33 was dated 11 September 1942, almost the same time as the first flight of the XB-32 at San Diego. The drawing is all that survived and no information is available as to the mission or engines, possibly the planned Pratt & Whitney R-4360s. It would appear that the B-32 airframe was relatively unchanged except for the incorporation of the B-29 single tail that had initially been planned for the B-32. (Convair via SDASM)

improved performance but the future direction of aircraft propulsion was clear with the development of gas turbine engines.

Consolidated had moderate success in the conversion of the B-24 into a land-based patrol bomber for the U.S. Navy, including the PB4Y-1 directly from the B-24 and the PB4Y-2 that had been heavily redesigned. A twin-engine version of the B-32 was offered in April 1945 with all of the added capabilities this larger aircraft could provide, possibly using the Pratt & Whitney R-4360 engine. This version included heavy defensive armament, five powered turrets, photo reconnaissance capability, and a suite of antisubmarine radar and countermeasure equipment.

There were commercial adaptations of the Model 33 from the earliest B-32 configurations but the first evidence, in February 1943, of a military transport is this straightforward version of the Model 33 with an 8-foot-6-inch-diameter cargo/passenger fuselage. It is believed there were probably earlier studies in parallel with the commercial studies that did not survive. (Convair via SDASM)

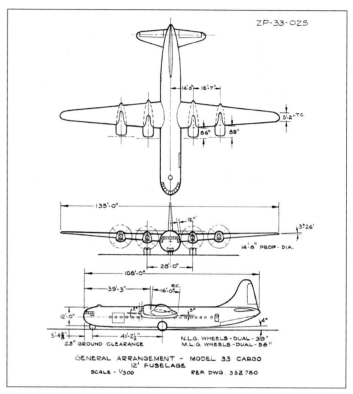

By 1943 the limitations of the B-32 fuselage and the attraction of developing a transport version led to a new fuselage design with increased diameter and length. This study configuration also appears to have the B-29 tail rather than the eventual Convair-designed vertical tail. The design had a rounded nose in the fashion of the XC-99 and had large forward-fuselage side cargo doors. (Convair via SDASM)

Large Model 33 fuselage is 12 feet in diameter and 108 feet in length, about 26 feet longer than the bomber. This transport version had the same wing, landing gear, R-3350 engines, and tail as the B-32. Gross weight of this transport was 100,000 pounds with an empty weight of 57,384 pounds. The range of this airplane was 4,400 miles with 10,000 pounds of cargo, or 1,350 miles with 130 troops aboard. (Convair via SDASM)

Transport conversions of the Model 33 were conducted in parallel with the basic studies, as was the practice with other programs. In this case, however, the first evidence available of a military transport version did not appear until 2 February 1943. It was essentially the XB-32 airframe, albeit with a single tail, and a fuselage configured for passengers and/or cargo. This may have been the reported para-troop transport conversion of the third B-32-20 built at Fort Worth.

The limitations of the B-32's fuselage for such a mission were apparent and in 1943 an entirely new fuselage was proposed to make a more attractive conversion. The new fuselage was enlarged to 12 feet in diameter and lengthened by 25 feet providing much improved accommodation for passengers and cargo. The

Bomber Programs

rest of the airframe was taken directly from the B-32 bomber design. The gross weight of this aircraft was 100,000 pounds and the empty weight was 57,400 pounds. It had a range of 4,400 miles with a payload of 10,000 pounds. The brochures proposing this transport illustrated accommodations showing a wide variety of cargo and tactical equipment and personnel provisions. A hospital MEDEVAC configuration would carry 91 litter patients, a troop transport of 130 troops, and when configured for paratroopers it would carry 92. The design still had the disadvantage of wing structure in the fuselage, effectively dividing the cabin into two compartments.

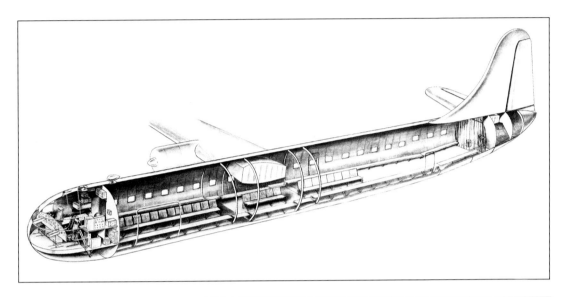

The illustration depicts the interior arrangement as a troop transport. There were accommodations for 130 troops, in this case, with bench seating at the fuselage sides and some central seats in the forward mid-fuselage. The mid-cargo compartment obstruction is readily apparent. There appears to be only one restroom in this design. (Convair via SDASM)

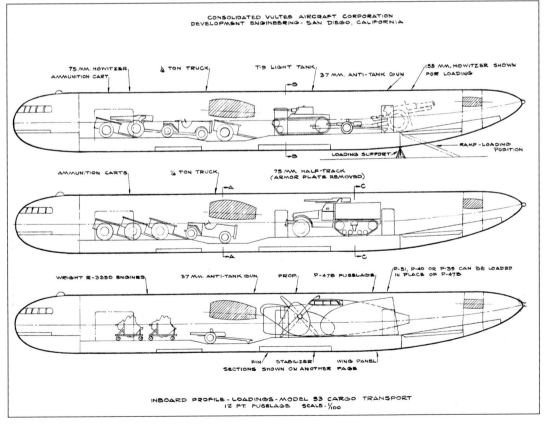

These views depict three of the various cargo loadings included in the brochures. The first carried a 75mm Howitzer, ammo cart, quarter-ton truck, T9 light tank, and 37mm anti-tank gun. Second loading included three ammo carts, a quarter-ton truck, and a 75mm half-track. The third was carrying a P-47 fighter aircraft and two R-3350 engines. Other loadings beside the previous 130-troop arrangement include 92 paratroopers and equipment, and a 91-litter hospital transport. (Convair via SDASM)

By December 1941, a commercial passenger configuration of the Model 33, designed somewhat earlier, was evolving in parallel with the XB-32. A somewhat less aerodynamic fuselage, the B-32 nose, and the twin tail with dihedral are apparent in this artist rendition. (Convair via SDASM)

Interior of the commercial version was also changed to a more spacious arrangement with seating for what appears to be 56 passengers. Entrance is at the wing and at the "low bridge" location, with some seating also at this location. A lavatory was in the tail and there appears to be two forward compartments seating eight each with their own lavatory compartment. (Convair via SDASM)

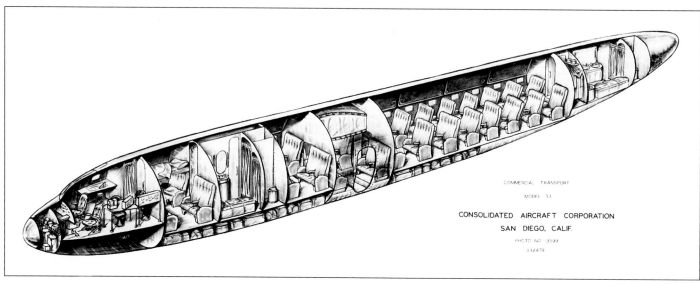

COMMERCIAL TRANSPORT
MODEL 33
CONSOLIDATED AIRCRAFT CORPORATION
SAN DIEGO, CALIF.
PHOTO NO. 3599
332479

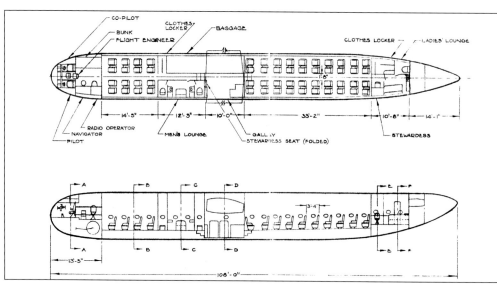

With the 12-foot-diameter fuselage design, a more useful interior arrangement was possible. It retained the previous 56-passenger capacity in two compartments, 16 forward and 40 in the aft fuselage. The area at the wing location was now used for the galley and baggage. Interestingly the lavatories are separated with a men's lounge forward and a women's lounge in the aft fuselage. Note also a bunk in the flight deck. (Convair via SDASM)

Bomber Programs

Commercial versions of the Model 33 were studied quite early, starting in mid 1941. The first example in August had a very aerodynamic fuselage reminiscent of the LB-25 but with a more B-32-looking nose and the trademark Consolidated vertical fins. The fuselage was lengthened significantly over the bomber and was configured to accommodate 78 day passengers or 34 at night in single berths. A crew of six included the two pilots, a radio operator, a navigator and two stewardesses. This airplane had a maximum range of 1,500 miles.

A commercial passenger version of the military transport configuration incorporating the 12-foot-diameter fuselage was studied in September 1943, not long after the military version was proposed. In this design, more spaciously configured for 58 passengers, 16 were located in the forward compartment and 42 in the aft compartment.

65 Model 35 (1940)

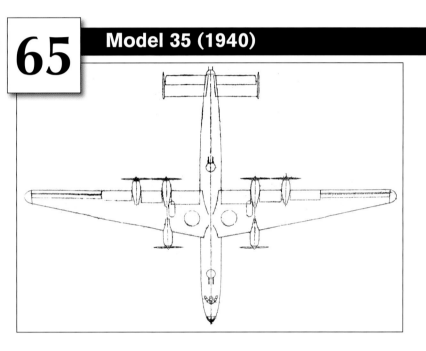

This Model 35 design is one of the only surviving drawings for a series of concepts Consolidated looked at during its intercontinental bomber studies in 1941. This particular configuration shows a four-engine (tractor-pusher) arrangement, with a wing area of 2,700 square feet and a wingspan of 164 feet 4 inches. Variants offered to the Army included this Model 35 configuration as well as several larger six-engine designs that eventually became the B-36. (Convair via SDASM)

In the later part of the prewar study period when the potential for the need for an ultra-long-range bomber was becoming apparent, the first study of a contemporary design for such a mission was undertaken. Design effort on what was to be designated as the Model 35 is believed to have started as early as September 1940 and culminated in the spring of 1941. The Model 35 turned out to be the immediate predecessor of the B-36 and included several of the design features that were apparent in the subsequent B-36 configurations including the circular fuselage, high wing, and twin tail.

Various layouts applicable to this requirement were investigated in the Model 35 study, including a four-engine design and three six-engine arrangements, and were offered in the initial B-36 proposal. The four-engine configuration featured an unusual pusher/tractor arrangement of the engines. This Model 35 version was obviously smaller than the six-engine designs and had a wing area of 2,700 square feet, a span of 164 feet 4 inches versus 4,772 square feet and 230 feet for the final B-36. The engines and the gross weight are not known but it obviously would not have the range or payload capability of the larger aircraft.

A similarly-configured design with four pusher and two tractor engines was included with the larger six-engine versions. It could be considered a logical growth of the smaller airplane. It is not clear if the efficiency of the pusher propeller operating in the propeller wash of the tractor engine would be affected or if other problems would occur. In any case, neither survived the competition. The remaining two designs had alternated aspect ratings and wing area. In addition, a six-engine tractor design was investigated, possibly at the same time.

All six-engine designs used basically the same airframes and were clear forerunners of the winner, which is readily recognizable as being close to the final B-36 design. The principal difference was the stepped cockpit design that was similar to the B-24 rather than that of the XB-36's that is similar to the B-32.

Display or wind tunnel model appears to be one of the Model 35 configurations with the long, forward under-wing engine air intakes and the more streamlined fuselage. The far-forward nose and the cockpit area are also consistent with the earliest design, as was the stepped B-24-type cockpit enclosure seen on this model. (Convair via SDASM)

66
B-36 Peacemaker - Model 36 (1941)

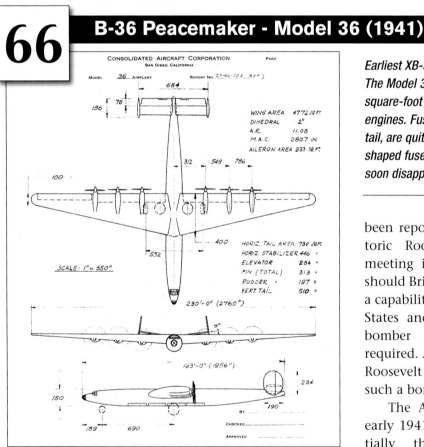

Earliest XB-36 configuration shows the legacy from the Model 35. The Model 36 was, of course, significantly larger, with a 4,772-square-foot wing versus 2,700 square feet, and it had six pusher engines. Fuselage lines, circular cross section, high wing, and twin tail, are quite similar. Initial design had a very aerodynamically shaped fuselage and a straight-wing trailing edge, both of which soon disappeared. (Convair via SDASM)

been reported that during discussions held at the historic Roosevelt and Churchill "Atlantic Charter" meeting in August 1941, it was acknowledged that, should Britain be invaded, the United States would need a capability that would allow operating from the United States and striking Germany and German forces. A bomber of intercontinental capability would be required. Apparently, as a result of these considerations, Roosevelt directed the Army Air Corps to start work on such a bomber.

The Air Corps had already initiated planning in early 1941 for airplanes of even greater capability. Initially the objectives were a much superior ultra-long-range bomber with a range of 12,000 miles, a bomb load of 4,000 pounds at the maximum range, and an average speed of 275 mph. These requirements, needless to say, were quite daunting at that time. On 11 April 1941, the Air Corps issued a Request for Proposal for a study of such an airplane from Boeing and Consolidated. The requirements called for a bomber capable

By 1940 the war was looking grim for the Allies, as most of Europe had fallen. The outcome of the Battle of Britain in the latter part of 1940 and early 1941 seemed to have thwarted Hitler's Operation Sealion plan for the invasion of Britain by denying Germany the air superiority that it would need. It has

Bomber Programs

The XB-36 prototype is shown flying over Carswell AFB on a test flight. The Convair plant is adjacent to this airfield, visible at the bottom of the photograph. First flight of the XB-36 took place on 8 August 1946 about four years and nine months after the original award in 1941. This overly long development period was due to conflicting priorities and manpower shortages primarily because of the mainstream war effort. (Convair via SDASM)

of 10,000 miles with a bomb load of 10,000 pounds at 250 to 300 mph. It should be capable of carrying a 72,000-pound payload for shorter distances and be capable of taking off from a 5,000-foot runway.

Consolidated had been studying large long-range aircraft with roughly similar capabilities and had the Model 35 study work on which to build. The company submitted several designs and variations including the six-engine pusher concept to the Army Air Corps under the Model 35 designation.

Boeing at the time was heavily involved in production of the B-17 and starting the B-29 program and did not put a lot of effort into this study, although it did respond. Two other companies also became somewhat involved. Douglas conducted some study but dropped out convinced that the requirements were beyond the capability of technology at that time. Northrop responded with the XB-35 flying wing concept, a program that interested the Air Corps and that ended up being developed in parallel with the B-36. Martin had

declined to participate in that it had little current study background in this type of requirement.

The study results for an intercontinental bomber were presented to Gen. "Hap" Arnold in October 1941 and Consolidated's six-engine pusher concept was recommended. Consolidated was subsequently awarded a contract on 15 November for the development and building of two XB-36 airplanes originally to be manufactured at San Diego. A new model number, Model 36, was assigned by Consolidated that worked well with the Air Corps' designation for the airplane. The B-36 program was now officially underway.

After the award in November 1941, the initial work on the program including the predesign and the mockup of the XB-36 were conducted at San Diego. In August 1942, however, the responsibility for the program was transferred to the Fort Worth Division because it had larger facilities and had become responsible for Consolidated's bomber business including the B-24 and the B-32. Only moderate changes to the proposed configuration

Bomber Programs

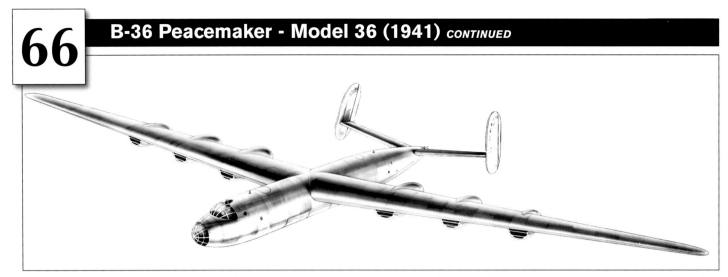

This view of the XB-36 in December 1941 illustrates, in the early design, the long air intakes for engine cooling, almost to the leading edge of the wing. At this point the constant-diameter fuselage had been adopted and the twin tail with horizontal stabilizer with pronounced dihedral had been retained. Nose shape, generally similar to the earlier B-32 treatment, is apparent. (Convair via SDASM)

were made. In February 1942, the cylindrical fuselage and rather blunt B-32-type nose replaced the initial more streamlined fuselage and nose. The twin tail was replaced with a single tail in January 1943 and the bubble-canopy flight deck was incorporated into the second prototype, the YB-36. The four-wheel main landing gear, replacing the large single wheel, first flew on the B-36A on 28 August 1947.

The B-36 program development was becoming bogged down in conflicting priorities with the B-24 and B-32 production, manpower shortages, and to a large extent, similar priority conflicts in the customer's shop, now the Army Air Force. In addition there were significant development delays of Pratt & Whitney's R-4360 engine. Because Consolidated only had a contract for two aircraft, there was difficulty in interesting subcontractors in the program. The Army Air Force did however issue a Letter of Intent for 100 ships in July 1943 and the contract was finalized in August 1944. Progress was quite slow but by 1945 the prototype was taking shape. The rollout of an incomplete XB-36, Number 1, occurred in September 1945 although much work remained for completion, modifications, and problem solving. The first engine runs took place on 12 June 1946 and, after a significant number of design changes, the first flight of the XB-36 finally took place on 8 August 1946.

The first production aircraft, the B-36A, was completed in July 1947 and flew on 28 August 1947. Operational deliveries to the Air Force started in June

1948. Overall the B-36 program had a very rough existence with much uncertainty, highly variable customer enthusiasm, varying support in Congress, and multiple threats of cancellation. As it turned out it was very fortuitous to have this weapon system as the prime nuclear strike force in the beginning and early phases of the Cold War with the Soviet Union.

The B-36 was named the Peacemaker as the result of a company contest in early 1949, but the name never received official sanction.

Several more advanced versions of the B-36 were offered by Fort Worth to the customer early in the program, including a B-36C, with six Pratt & Whitney R-4360 Variable Discharge Turbine (VDT) engines arranged in a tractor design. Another featured four Wright T35-1 or -4 turboprops, again in a tractor engine arrangement. With the T35-4, the B-36 had a top speed of 417 mph, a gain of 58 mph more than the B-36B. Two B-36s were also converted to a swept-wing all-jet version with eight Pratt & Whitney J57 turbojets as a competitor to the B-52. None of these concepts were continued.

An upgrade of the B-36 was carried out when jet engine pods, with two J35s each, were added to the wingtips. This increased the performance significantly and effectively forestalled competitive systems until the B-52 all-jet bomber program was initiated. The upgrade modification program to add these jet pods was carried out at San Diego.

Bomber Programs

CONSOLIDATED MODEL 36 TRANSPORT

CONSOLIDATED AIRCRAFT CORPORATION
SAN DIEGO, CALIF.

Transport version of the XB-36 (Model 36) intercontinental bomber was studied as early as mid 1941 and was proposed to the Army in August. These first concepts were based on the initial XB-36 that had a rather streamlined fuselage and twin tails. Gross weight was 265,000 pounds and it could carry 250 troops a distance of 3,475 miles at an average speed of 300 mph. (Convair via SDASM)

Consolidated was never hesitant about looking at transport and commercial passenger versions of new military designs that were being studied. It was no different when initial work was underway for the latest heavy intercontinental bomber that was to become the XB-36 (Model 36). Both military transport and passenger concept adaptations of the basic bomber design were initiated as early as mid 1941 and Laddon recommended the development of such an airplane to Army customers in August 1941. The Army ordered one XC-99 on 31 December 1941 that was scheduled for delivery in 21 months on a non-interference basis with the XB-36. The first available brochure for this six-engine transport was published 30 January 1942, just 10 weeks after the contract was signed for the XB-36 bomber aircraft. This project, as well as the XB-36 itself, was much delayed by a combination of low priorities and lack of personnel engendered by the mainstream war effort.

Earliest versions of this transport were based on the original slimmer, more streamlined Model 36 design with twin tails. As the configuration evolved, the fuselage for the transport version increased from the bomber's diameter of 12 feet 6 inches to that of a two-deck pressurized fuselage with an elliptical cross section 13 feet 6 inches wide by 19 feet 4 inches high. The wing, powerplants, landing gear, and tail surfaces would be used directly from the B-36 design and carried a crew of five. Maximum speed was quoted at 345 mph. The cargo version carried a basic cargo weight of 68,000 pounds for a range of 3,880 miles.

As the XB-36 design evolved, those changes were incorporated in the transport concepts, including the change to a single tail. By August 1942 the XC-99's fuselage had been further enlarged to a cross section of 14 feet 3 inches wide by 20 feet 6 inches high and it was about 10 feet longer. It had a revised, more-blunt nose, but was still somewhat more streamlined compared with the final design. The capacity of this version had been increased to be able to carry 400 troops with a range of 3,250 miles at an average speed of 210 mph. It was also capable of transporting 100,000 pounds of cargo for shorter distances.

The first available documentation of the final design iteration of the XC-99, now the Model 37, was published in June 1943. Only a few changes were noted at this point, mainly the more-blunt, shortened nose that would be seen on the aircraft as it was eventually built. It now had a crew of 10. The specifications and capabilities generally remained unchanged. The gross weight remained as planned and the maximum range with 10,000 pounds of payload was 8,100 miles. The range with 100,000 pounds of payload was 1,720 miles.

As originally planned, the wing, engines, landing gear, and tail services were provided from Fort Worth's B-36 production line, and the fuselage itself was built in two sections at San Diego. The aircraft final assembly also took place at San Diego. The XC-99 was finally ready for its maiden flight, which took place on 24 November 1947. After the initial flight testing, an upgrading modification program was undertaken before delivery to the Military Air Transport Service (MATS) on

Bomber Programs

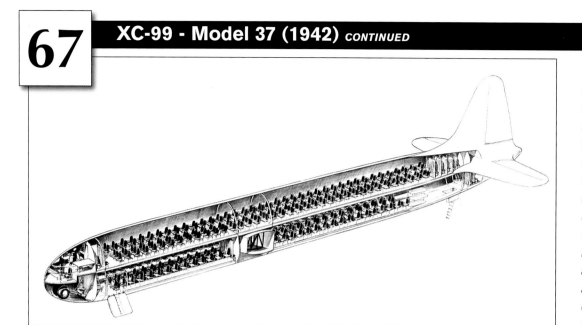

This troop transport version carried 400 passengers on two decks. There were two loading entrances, one forward, and one at the rear of the airplane. Lavatories were included on both decks at the front and rear of the fuselage. It appears there was one stairway between decks forward just aft of the flight deck and another one at the far rear of the fuselage. (Convair via SDASM)

New version of the transport offered revised bottom-loading doors both fore and aft. In this case the rear door/loading ramp is shown deployed in a rather steep incline. The 3-inch anti-aircraft gun is apparently being winched up the ramp with the tractor truck. The concept and execution in this depiction seems to be rather primitive compared to the requirements of the day. (Convair via SDASM)

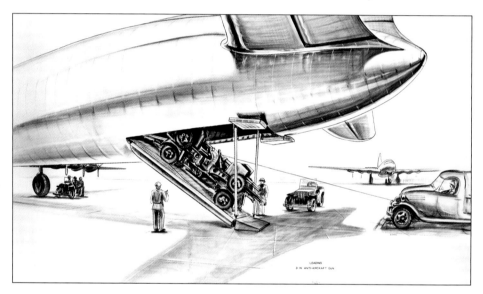

26 May 1949. It was noted in the Air Force's Final Report on the development and testing of this aircraft that, although the original needs were valid, the overall logistic aspects had changed considerably. Specifically there was need for bulkier cargo and the capability for more concentrated loads. The XC-99's shortcomings included the restricted internal dimensions, the floor line that was too high, straight-away loading that was too restricted, and tie-down fittings that were inadequate. The XC-99 was employed in limited operation until its retirement eight years later in March 1957.

An advanced version of this aircraft was proposed in 1949, apparently in a competition for the heavy-cargo requirement that was eventually fulfilled by the Douglas C-124 Globemaster. This variant of the C-99 was to have a somewhat enlarged fuselage (15 feet 8 inches wide and 22 feet 7 inches high), more powerful engines, and new clamshell doors in the nose and tail. A standard B-36 flight deck was also to have been incorporated in this follow-on proposal. There are some reports that turbo-props, T34s or T56s, had also been considered for follow-on versions.

Commercial versions were also studied early in the B-36 program and a passenger version was shown in artwork of the airplane in civilian dress very reminiscent of the PAA color scheme. Little information was available with regard to the specifications for this initial civilian version but it was generally similar to the military

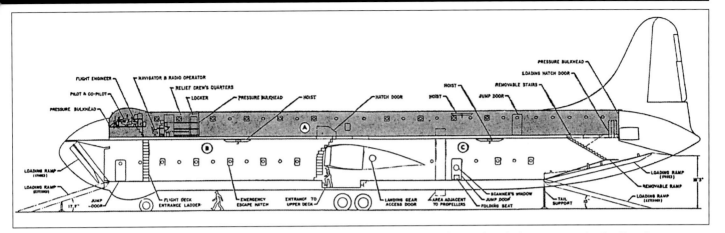

An advanced version of the XC-99 was offered in the competition in 1949 for a Heavy Cargo aircraft that was won by the Douglas C-124 Globemaster. XC-99 shown here was to have had much-improved loading and unloading capability with larger nose and tail clamshell doors, and incorporated the B-36 flight deck design for the crew. It was also to have more powerful engines, and turboprops were considered. (Convair via SDASM)

Although the XC-99 was contracted at the end of 1941, it and the B-36 itself, were much delayed in the development process. Causes of this delay were primarily the lack of customer priority and the lack of engineering and manufacturing personnel due to the war effort. First flight of the XC-99 took place on 24 November 1947, about 16 months after the XB-36. It is shown here on a test flight in the San Diego area. (Convair via SDASM)

design. It was to accommodate 144 day passengers convertible to 68 sleeping berths at night.

As the war was winding down, PAA was definitely interested in the Model 37 for both the European and Hawaii passenger markets. In February 1945 PAA ordered 15 of these Super Clippers, 12 for the Atlantic runs and three for the Hawaii route. The commercial specification, dated 15 February 1945, called for a 204-passenger aircraft with a maximum gross weight of 320,000 pounds.

The maximum speed was quoted at 370 mph at 20,000 feet altitude. It is believed that the aircraft was originally planned to have turboprop engines but their delayed development and questionable availability necessitated a return to the standard Pratt & Whitney R-4360 powerplant. Pan American Airways, after second thoughts and a reanalysis of the economics involved, concluded that this was not the way to go and canceled the order late in 1945.

Illustration depicts passenger loading through a forward door as well as an aft passenger ramp. This heavily retouched model photo shows the Model 37 in generally similar markings as Pan American World Airways was using at the time. The illustration was probably released either prior to or shortly after PAA ordered 15 of these aircraft in February 1945. PAA, after a market reevaluation, canceled the order by the end of 1945. (Convair via SDASM)

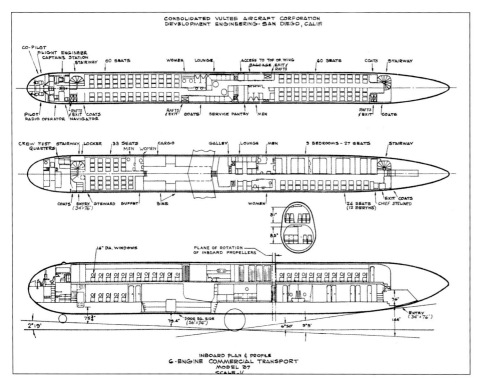

Interior arrangement of the Model 37, as sold to PAA, is believed to be depicted in the Outline Specification dated 15 February 1945. This interior accommodated 120 on the top deck, 57 on the lower deck, and 27 in nine bedrooms also on the lower deck, for a total of 204 passengers. Although the Model 37 would have been a monster airliner in the late 1940s the three and two seating is very familiar by today's standards and the total of 204 passengers would be quite modest for such a large modern-day aircraft. (Convair via SDASM)

The long and short of it in 1948: Convair's mammoth six-engine XC-99 Cargo transport prototype parked next to the company's single-engine L-13 light observation aircraft nicely illustrates the range and scope of company products. Less than a decade later, Convair's product line included bombers, seaplanes, airliners, fighters, transports, and guided missiles. (National Archives via Dennis R. Jenkins)

Bomber Programs

Appearing quite primitive when compared to today's digital "glass cockpit"-equipped aircraft, the XC-99's flight deck was state of the art when the giant cargo transport first flew in 1947. Outward visibility was quite excellent and the layout of all controls and instrumentation was exemplary from a pilot ergonomics point of view. Flight Engineer occupied the center seat. (National Archives via Dennis R. Jenkins)

68 Flying Wing B-36 Comparison (1942)

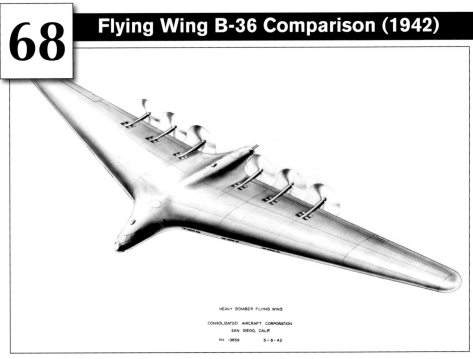

HEAVY BOMBER FLYING WING

CONSOLIDATED AIRCRAFT CORPORATION
SAN DIEGO, CALIF.

PH 13659   5-8-42

At the time of the B-36 contract award, Northrop had interested the Army in a flying-wing concept it had been developing for several years. Northrop received a contract for the XB-35 that was to run almost in parallel with the B-36. Consolidated, either at the request of the Army or in its own self-interest, conducted a competitive evaluation in mid 1942 of a large six-engine flying wing using the identical requirements as the B-36. This resulting flying-wing design had a wing area of 7,500 square feet, a wingspan of 288 feet, and a length of 78 feet. This compares to the B-36's 2,772-square-foot wing and 230-foot wingspan. The wing's gross weight was 237,800 pounds—

significantly less than the B-36's 328,000 pounds. The flying wing was powered by the same Pratt & Whitney R-4360 engines as the B-36 that were also used in a pusher configuration. (Convair via SDASM)

Bomber Programs

The origins of the B-36 lie in the Army Air Corps solicitation to Boeing and Consolidated on 11 April 1941 for a long-range heavy bomber capable of carrying a 10,000-pound bomb load and having a range of 10,000 miles. Boeing and Consolidated responded, as did Douglas and Northrop with unsolicited proposals.Consolidated was the winner. The Northrop proposal for a flying-wing design, however, elicited sufficient interest that they were provided a separate prototype contract on 22 November 1941, shortly

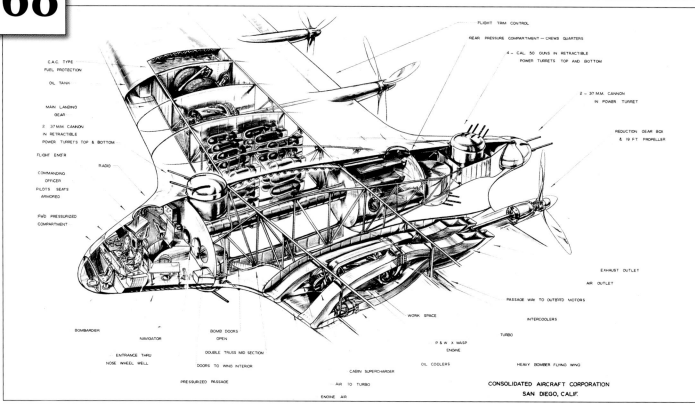

FLIGHT TRIM CONTROL
REAR PRESSURE COMPARTMENT — CREWS QUARTERS
4 - CAL. 50 GUNS IN RETRACTIBLE
POWER TURRETS TOP AND BOTTOM
C.A.C. TYPE
FUEL PROTECTION
OIL TANK
2 - 37 M.M. CANNON
IN POWER TURRET
MAIN LANDING
GEAR
REDUCTION GEAR BOX
& 19 FT PROPELLER
2 37 M.M. CANNON
IN RETRACTIBLE
POWER TURRETS TOP & BOTTOM
FLIGHT ENG'R
RADIO
COMMANDING
OFFICER
PILOTS SEATS
ARMORED
FWD PRESSURIZED
COMPARTMENT
EXHAUST OUTLET
AIR OUTLET
PASSAGE WAY TO OUTB'R'D MOTORS
WORK SPACE
INTERCOOLERS
BOMBARDIER
TURBO
NAVIGATOR
BOMB DOORS
OPEN
P & W X WASP
ENGINE
ENTRANCE THRU
NOSE WHEEL WELL
DOUBLE TRUSS MID SECTION
OIL COOLERS
HEAVY BOMBER FLYING WING
DOORS TO WING INTERIOR
PRESSURIZED PASSAGE
CABIN SUPERCHARGER
AIR TO TURBO
CONSOLIDATED AIRCRAFT CORPORATION
SAN DIEGO, CALIF.
ENGINE AIR

Interior arrangement of the evaluation design is shown in this illustration. There were crew quarters at the aft end on the wing that also included a retractable four .50-cal. top gun turret and a twin 37mm tail turret. Defensive armament also included front, top, and bottom twin 37mm gun turrets. The main flight crew compartment was in the nose of the aircraft while the engines were located forward in the wing with long shafts driving the pusher propellers. (Convair via SDASM)

The brainchild of design impresario John K. Northrop, the XB-35 Flying Wing was years ahead of its time, and became a direct competitor to Convair's B-36 Strategic Bomber. Although an exceedingly clean design aerodynamically, inherent instability and the unavailability of modern, digital flight-control systems relegated the Flying Wing to less-than-favorable status as a military aircraft. The design hopes of Northrop and Convair with their flying-wing bombers were not realized until half-a-century later with the entry into service of Northrop's B-2 Spirit Stealth Bomber. Wingspan of the XB-35 and B-2 is an identical 172 feet 0 inches. (National Archives via Dennis R. Jenkins)

after the contract was issued for the XB-36. The development for these two programs ran in parallel, delays and all, with the first flights within two months of each other in 1946.

In 1942 Consolidated conducted a design study to compare an all-wing configuration to that of the B-36. It is not clear if this was at the request of the customer or if it was competitively motivated to understand the XB-35 flying-wing design approach.

Consolidated had very little experience with all wing designs compared to Northrop, which had flown its N1M in July 1940 and had conducted much design

Bomber Programs

and research work on the all-wing concept. Consolidated's concept of a "flying wing" in 1937 had a conventional tail, but carried the entire useful load in the wing. Because of the need for expertise with regard to a true flying wing, it is believed that Consolidated engaged a consultant, Wilhelm "Bill" F. Schult, who had worked extensively on such designs for Ryan, Lockheed, and Northrop. He had a long career interest, from the 1930s, in the flying-wing concept and built many wind tunnel models that he apparently tested independently. Although there is only a small amount of direct evidence available, it is believed that the general design of this study airplane was based on his work. He did provide basic design work to Consolidated based on GALCIT wind tunnel data on possibly his own model design.

The study objective was to determine the characteristics of a six-engine flying wing capable of a range of 10,000 miles that could carry a 10,000-pound payload. Propulsion was the same as the B-36's six pusher Pratt & Whitney R-4360 Wasp Major engines. Its maximum speed was 394 mph compared with 378 for the B-36, but the cruise speed was 60 mph slower than the B-36 at 210 mph.

The conclusion from the study was that the flying-wing concept looked very promising. Its lower cruise speed was due to the lower wing loading and the increase in landing distance of 1,000 feet was not objectionable because it was still less than the B-36 takeoff distance. The big unsolved problem, it was noted, was that of providing adequate control. It was felt that split ailerons could provide satisfactory lateral and directional control but longitudinal control may have been more difficult. The intent was to test a model, then being built in May 1942, in the GALCIT wind tunnel to further explore its aerodynamics.

69 | PB4Y Privateer - Model 100 (1942)

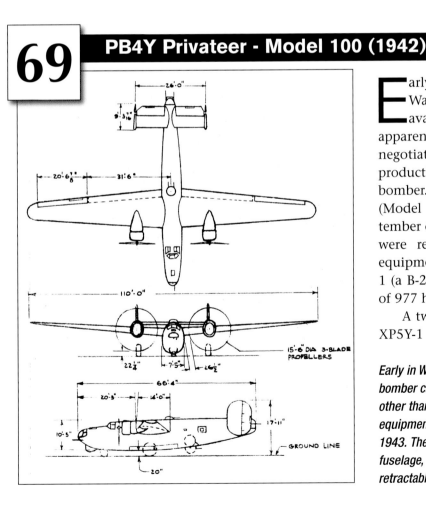

Early on in the United States' involvement in World War II in the Pacific, the vulnerability of the Navy's available seaplane patrol aircraft was readily apparent. In July 1942 , the Navy was finally able to negotiate with the Army for a share of the B-24 bomber production for use as a long-range land-based patrol bomber. These B-24Ds were redesignated the PB4Y-1 (Model 100) and the first Navy aircraft flew on 7 September of that year. These B-24s to be used by the Navy were relatively unmodified aircraft other than the equipment needed for the Navy mission. The last PB4Y-1 (a B-24M) was delivered in January 1945 after a total of 977 had been built.

A twin-engine B-24 was offered to the Navy as the XP5Y-1 (Model 38) Long-range Patrol Airplane in

Early in World War II, the Navy used the B-24 as a maritime patrol bomber called the PB4Y-1, which was essentially unchanged other than incorporating the appropriate naval mission equipment. A redesigned version, the PB4Y-2, first flew in May 1943. The modified aircraft included a lengthened forward fuselage, the new single tail, and added gun turrets and retractable radar. (Convair via SDASM)

October 1942 but was turned down by the Navy. This aircraft was a basic B-24D that had the four R-1830 engines replaced by two Wright R-3350 engines. It had a design gross weight of 50,000 pounds and a nine-man crew. As an example, for a mission carrying eight 325-pound depth bombs, it had an endurance of 17.3 hours and a range of 2,450 smi cruising at 170 mph at 5,000 feet. The Navy's rejection cited the similarity to the PB4Y-1 already underway, and the shortage and delayed production of the R-3350 engine. The Navy did agree that Convair should continue study on the two-engine B-32.

A redesigned version of the PB4Y-1, optimized for the Navy low-altitude patrol mission, was ordered in May 1943, and three PB4Y-1s were allocated for prototype conversion. The modifications for this version were more extensive and included the lengthening of the fuselage 7 feet forward of the wing, a new single tail, removal of the engine turbo superchargers, new nacelles, added defensive armament, and retractable radar. This armament consisted of a Convair nose and tail turret, two Martin top turrets, and two ERCO fuselage side turrets. The wing and the landing gear were unchanged. It carried a crew of 11.

The initial flight of the first converted PB4Y-2 occurred on 20 September 1943. A contract for 600 was awarded in October of that year and the first flight of the production aircraft took place on 15 February 1944. A total of 739 aircraft were finally produced and the airplane was in service until 1954.

After the Navy developed the improved PB4Y-2, it also ordered transport versions of that aircraft designated RY-3 (Model 101). This version of the Privateer was capable of carrying 28 passengers. A large quantity was ordered but only 34 were actually built and most were allocated to the British.

The U.S. Navy was offered a twin-engine B-24, called a P5Y, as a Long-range Patrol Airplane in October 1942, but it was rejected because it was too similar to the ongoing PB4Y-1 and because of the unavailability of the proposed R-3350 engines. Airframe was identical to the B-24D except for the engine installation and the accommodation of the Navy's mission equipment. (Convair via SDASM)

Bomber Programs

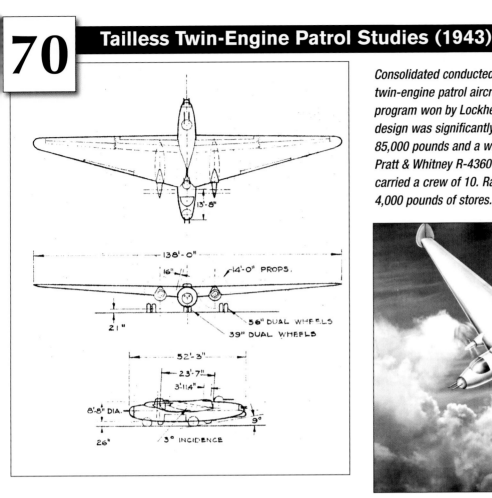

Consolidated conducted studies and possibly a proposal for a twin-engine patrol aircraft in August 1943 believed to be for the program won by Lockheed with its P2V Neptune. This tailless design was significantly larger than the P2V with a gross weight of 85,000 pounds and a wingspan of 138 feet. It was powered by two Pratt & Whitney R-4360 engines with 3,250 hp for takeoff, and carried a crew of 10. Range was 4,000 miles with a payload of 4,000 pounds of stores. (Convair via SDASM)

Configuration was revised in December 1943 with changes that included reducing the wing area to 1,800 square feet, increasing the gross weight to 90,000 pounds, adding wingtip-mounted vertical-control surfaces, and various other structural and landing-gear changes. These changes extended the aircraft's range to 5,000 miles and increased the top speed to 288 mph at sea level. (Convair via SDASM)

After the B-36 program was underway in 1942 and Convair had conducted a comparison study of the conventional aircraft configuration and the flying-wing concept being promoted by Northrop, Convair became interested in the tailless concept. This aircraft design, developed in the latter part of 1943 for Navy patrol and Army bomber applications, was again not a true tailless craft but used extensible horizontal tail surfaces for takeoff and landing.

In 1943 the Navy was actively seeking a replacement upgrade for the Lockheed PV-1 Ventura and PV-2 Harpoon widely used in World War II. Since 1941, Lockheed had been studying upgrades to these aircraft that would provide a longer range and increased payload, and the Navy was reported to have given Lockheed a Letter of Intent for a new patrol aircraft in February 1943. The program apparently did not have the highest priority but a contract for the XP2V-1 Neptune was awarded to Lockheed on 4 February 1944.

In this context it seems that Convair's tailless concept was aimed at the PV replacement program. In fact it is interesting that several of the graphics associated with this proposal use the designation P5Y in the title, but it is doubtful that the designation had any official standing and reflects the rather undisciplined use of designations at that time.

Convair's initial studies of a tailless twin-engine patrol airplane and a four-engine bomber were completed in August 1943. As mentioned, these designs used an extensible horizontal stabilizer and elevator to alleviate perceived low-speed stability and control issues associated with flying-wing concepts at the time. This configuration was a fairly large aircraft, certainly much larger than the P2V. It had a crew of 10 and a range of 4,000 miles while carrying a standard payload

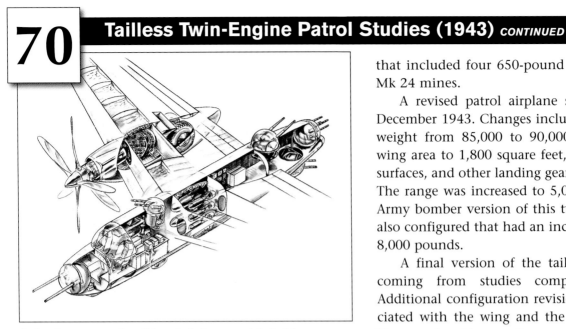

Defensive armament on this patrol plane consisted of five gun turrets including a nose and tail position with twin 20mm cannons. A ventral rear turret and a fore and aft top fuselage turret all had twin .50-cal. guns. Aft fuselage turrets were retractable. Fuselage bomb bay located at the wing is shown with the basic four 650-pound depth charges and two Mk 24 mines. The buried R-4360 engine installation is shown driving the coaxial counter-rotating 16-foot props via an extension shaft. (Convair via SDASM)

that included four 650-pound depth charges and two Mk 24 mines.

A revised patrol airplane study was published in December 1943. Changes included increasing the gross weight from 85,000 to 90,000 pounds, reducing the wing area to 1,800 square feet, adding wingtip vertical surfaces, and other landing gear and structural changes. The range was increased to 5,000 miles as a result. An Army bomber version of this twin-engine airplane was also configured that had an increased bomb capacity of 8,000 pounds.

A final version of the tailless concept was forthcoming from studies completed in April 1944. Additional configuration revisions were primarily associated with the wing and the extensible tail surfaces. The aspect and taper ratios were changed to improve the stall characteristics, the wing thickness was decreased, and the wing was moved to the mid position on the fuselage. An increased extension of the revised rear surfaces was also adopted. Since the contract for the Lockheed P2V was awarded in February 1944, that requirement appears to have been fulfilled and no further work was carried out on the tailless concept.

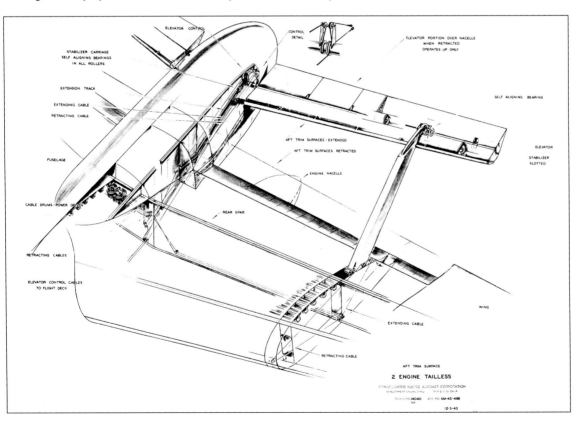

Structural arrangement of the extensible horizontal stabilizer and elevator are detailed in this drawing. These aft surfaces moved along a track at the fuselage and were supported in the extended position by a large retractable strut outboard of the fuselage. These surfaces were only extended during low-speed flight, and when retracted, became integral with the wing. (Convair via SDASM)

Bomber Programs

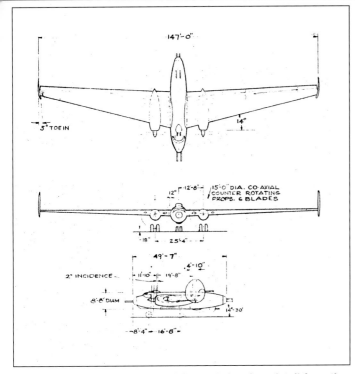

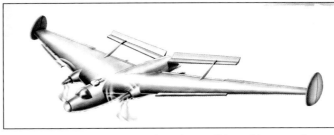

Revised design is shown in this artist rendition with some changes in the tail extension structure in the low-speed landing configuration. This view shows the absence of one of the top turrets, one of the alternate versions studied. (Convair via SDASM)

Revision of the patrol airplane differs only in minor detail from the original design of a year earlier. In summary, the wing was reconfigured and wingtip vertical surfaces added. Engines were moved from a buried position to a location ahead of the main spar but, otherwise, the configurations are very similar. This was the final version and no further work was undertaken on these tailless designs. (Convair via SDASM)

Further revised design was completed in April 1943 that changed the wing's aspect and taper ratios and placed the engines ahead of the spar and in nacelles to improve the stall characteristics. Wing was moved to a mid-fuselage position, made thinner, and the nose was shortened. The rear control surfaces were also redesigned to have a greater extension. (Convair via SDASM)

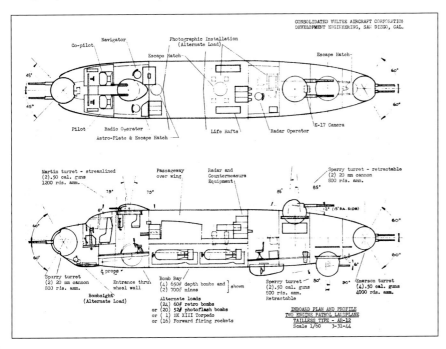

Inboard profile drawing depicts the revised bomb-bay configuration and the location of the defensive gun turrets. Nominal payload used when quoting performance was four 650-pound depth charges and two 700-pound Mk 24 mines. Alternate payloads, such as photoflash bombs for photoreconnaissance missions and torpedoes were also quoted. (Convair via SDASM)

A conventional patrol airplane was designed to provide a comparison with the tailless concept proposed several months earlier. In this study, tactical mission and equipment requirements were equalized so that the resulting differences in size and performance would be solely due to the differing aerodynamic designs being compared. This configuration had the same two Pratt & Whitney R-4460 engines, crew arrangement, defensive armament, and range requirement. (Convair via SDASM)

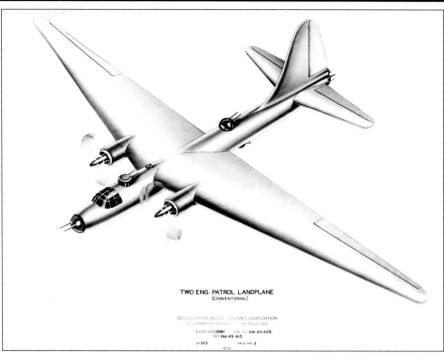

TWO ENG. PATROL LANDPLANE
(CONVENTIONAL)

CONSOLIDATED VULTEE AIRCRAFT CORPORATION
DEVELOPMENT ENGINEERING   SAN DIEGO, CALIF.
PHOTO NO.1396I   DES. NO XM-43-429
REF. XM-43-415
ZP-013   PAGE NO 3
DATE

CONSOLIDATED VULTEE AIRCRAFT CORP.
DEVELOPMENT ENGINEERING - SAN DIEGO, CALIF.
PAGE 6
ZP-013
52'-4"
152'-0"
14'-2" PROP
1'-11"
87'-4"
20'-3"
24'-0"
4'-0"
8'-2"
2'-8"
6"

The resulting aircraft had a wingspan of 152 feet, and a gross weight of 89,500 pounds, an increase of 14 feet (10 percent) and 4,500 pounds (5.3 percent), respectively. The range was held the same for the two aircraft, at 4,770 miles with no bombs, as was the takeoff distance. Powerplants were the same also, which resulted in a larger, heavier vehicle that was consequently slower by 35 mph (14 percent) at 252 mph (military power at sea level). (Convair via SDASM)

In October 1943, two months after the initial report on the tailless design in August 1943, a study was published that compared the tailless concept with conventional designs for both the tailless and the four-engine version, to highlight the potential advantages that this new approach might provide. The results of those comparisons indicated the tailless design was somewhat smaller and lighter for the same mission requirements. Its advantage stemmed primarily from reduced drag due to the elimination of the tail, a much shorter fuselage, and the elimination of the engine nacelles. In the case of the two-engine patrol aircraft, the wing area was 1,900 square feet for the tailless versus 2,100 square feet for the conventional design, and the gross weight of the conventional design was 4,500 pounds heavier at 89,500 pounds.

Bomber Programs

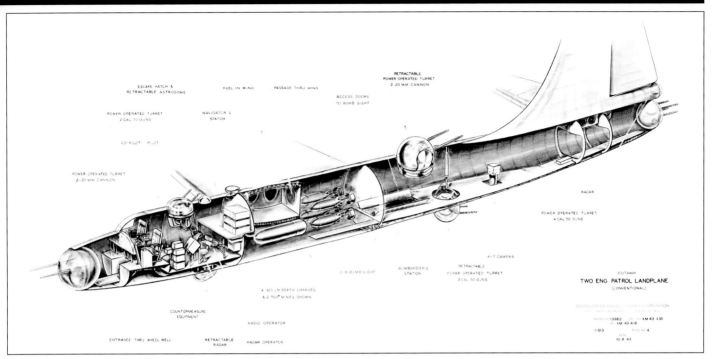

Interior arrangement was generally similar to the tailless design although the fuselage obviously had more interior volume available. Two retractable fuselage gun turrets were at mid fuselage, a twin .50-cal. turret was immediately behind the flight deck, and there was a nose and a tail turret. Bombardier was located in a unique position just behind the bomb bay and in front of the fuselage turrets, necessitated by the nose turret installation. *(Convair via SDASM)*

72 Tailless Four-Engine Army Bomber (1943)

Four-engine tailless bomber was a direct scale-up of the twin-engine design and was visually very similar to the original patrol design. This design was configured as a bomber for the Army that was capable of carrying up to 40,000 pounds of bombs and had a range of 7,500 miles with a 5,000-pound bomb load. It was powered by four Pratt & Whitney R-4360 Wasp Major engines and had a gross weight of 180,000 pounds with a crew of nine. (Convair via SDASM)

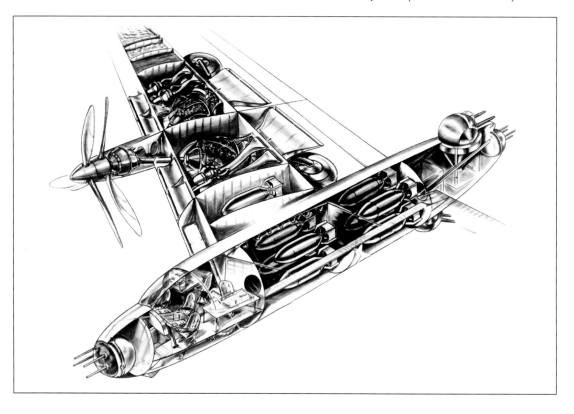

```
           ├── 43'-3" ──┤
                12'-9"
              ├─19'-4"─┤
```

```
      ├─────────── 190'-0" ───────────┤
20'-0" DIA.
COUNTER ROTATING
                          20" GROUND CLEARANCE
        56" DIA. S.C. DUAL WHEELS
  66" DIA S.C. DUAL WHEELS
```

```
        ├──── 62'-0" ────┤
        ├── 34'-4.3" ──┤
          2° INCIDENCE
            5'-5"
  10'-6"                      9°
  3'10"  ├─10'-6"─┤─ 24'-0" ─┤
```

GENERAL ARRANGEMENT -
4 ENGINE TAILLESS BOMBARDMENT AIRPLANE
SCALE - 1" = 30'

A t the same time as the twin-engine patrol study was done, a four-engine Army bomber was also designed. This configuration was significantly larger, about twice the size of the patrol airplane, with a wing area of 3,600 square feet and a gross weight of 180,000 pounds, and was powered by four Pratt & Whitney R-4360 engines. These engines, buried in the wing, were arranged in two pairs with each pair driving 20-foot counter-rotating propellers. This bomber had a range of 7,500 miles with an 8,000-pound bomb load. The configuration of this bomber is virtually identical to a scale-up of the twin-engine patrol aircraft. The four-engine conventional design was 20,000 pounds heavier and the wing area was larger by 24 percent at 4,450 square feet. After this brief study no further work was carried out on this concept.

The four-engine Army bomber had a wing area of 3,600 square feet and a span of 190 feet. Its fuselage was 62 feet long. Retractable tail surfaces had a span of 86 feet 6 inches. Pratt & Whitney R-4360s were installed in pairs, driving the two counter-rotating propellers. These engines were rated at 3,250 hp military power. (Convair via SDASM)

Buried engines were installed in angular mounted pairs, driving three-blade counter-rotating propellers that had a diameter of 20 feet. The gun positions included a nose and a tail turret with four .50-cal. guns and a ventral and dorsal fuselage turret housing twin 37mm cannons. The crew was accommodated in the forward and rear fuselage compartments. (Convair via SDASM)

Bomber Programs

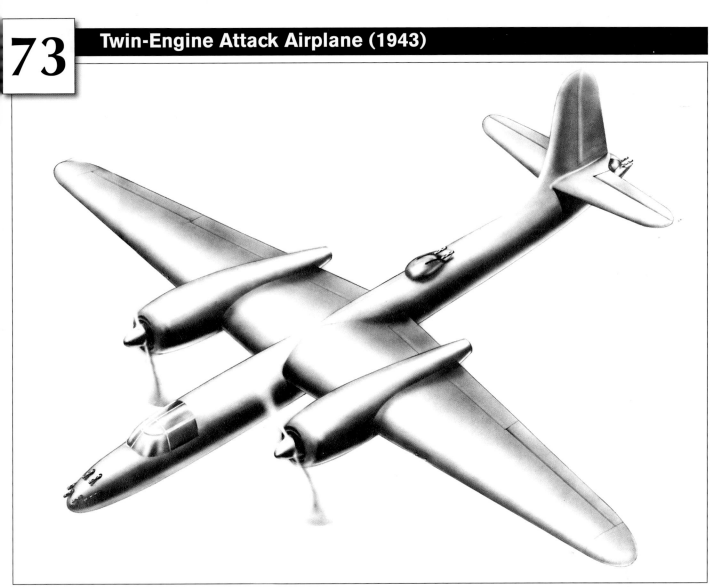

A design for a twin-engine attack and light bomber was developed in July 1943. This airplane had a gross weight of 35,000 pounds and carried a crew of four. Engines were Pratt & Whitney R-4360 Wasp Majors delivering 3,250 hp military power. This design may have been in response to Army interest in a ground-attack airplane similar to the smaller XA-38 prototype ordered earlier. That aircraft, as well as the B-25, carried a nose-mounted 75mm gun. (Convair via SDASM)

A light twin-engine attack airplane was configured and a brochure published in July 1943. A possible motivation for this design was the Army's interest in the Beech XA-38 with the 75mm nose cannon for attack missions, although a prototype contract for that airplane had been awarded in December 1942.

The twin-engine attack airplane was a conventional configuration with two Wasp Major engines of 3,250 hp and a crew of four. An augmented exhaust system was incorporated and used to manipulate the engine exhaust in such a manner as to provide a jet-propulsion assist effect. It had a gross weight of 35,000 pounds, a maximum bomb load of 4,000 pounds, and a maximum speed of 397 mph.

Defensive armament included a twin .50-cal. tail turret, a twin .50-cal. Martin fuselage turret, and provision for a single flexible .50-cal. gun in the bottom of the aft fuselage. Forward firing attack armament included four wing-mounted .50-cal. guns and a nose installation with three armament alternates. These were six .50-cal. guns, or three 37mm cannons, or a single 75mm cannon.

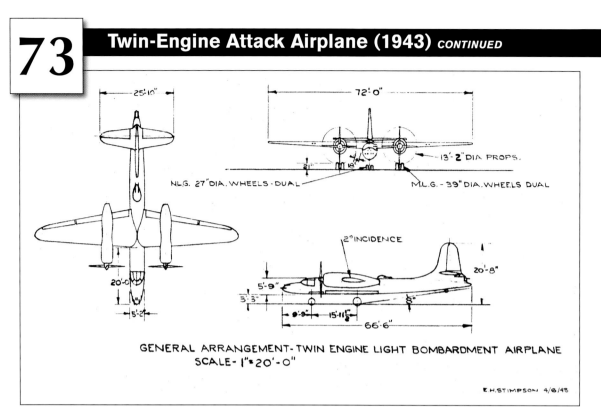

25'-10"

72'-0"

13'-2" DIA PROPS.

NLG. 27" DIA. WHEELS - DUAL

M.L.G. - 39" DIA. WHEELS DUAL

21"

18"

2° INCIDENCE

20'-8"

5'-9"

3'-3"

8°

9'-9"

15'-11½"

66'-6"

20'-0"

5'-2"

GENERAL ARRANGEMENT- TWIN ENGINE LIGHT BOMBARDMENT AIRPLANE
SCALE- 1"=20'-0"

E.H.STIMPSON 4/8/43

This airplane had a 650-square-foot wing with a 72-foot span. The fuselage was 66 feet 6 inches long. The large engines provided a top speed of 400 mph, 387 mph at military power. It had a range of 1,200 statute miles carrying 2,000 pounds of bombs. *(Convair via SDASM)*

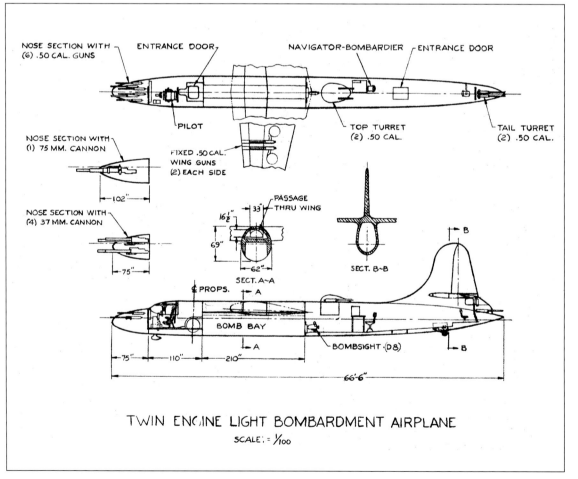

NOSE SECTION WITH (6) .50 CAL. GUNS

ENTRANCE DOOR

NAVIGATOR-BOMBARDIER

ENTRANCE DOOR

PILOT

NOSE SECTION WITH (1) 75 MM. CANNON

FIXED .50 CAL. WING GUNS (2) EACH SIDE

TOP TURRET (2) .50 CAL.

TAIL TURRET (2) .50 CAL.

102"

NOSE SECTION WITH (4) 37 MM. CANNON

PASSAGE THRU WING

33"

16½"

69"

62"

SECT. A-A

75"

SECT. B-B

C PROPS.

A

B

BOMB BAY

A

BOMBSIGHT (D8)

B

75"

110"

210"

66'-6"

TWIN ENGINE LIGHT BOMBARDMENT AIRPLANE
SCALE = 1/100

Defensive armament included a tail turret and a Martin fuselage top turret, both armed with twin .50-cal. guns. Ground-attack armament, in addition to its 4,000-pound payload capability, consisted of four .50-cal. guns in the wing and three alternate nose configurations. The nose could accommodate six .50-cal. guns or three 37mm cannons or a single 75mm cannon. The bomb bay was directly under the wing and the pilot was the only occupant of the flight deck. *(Convair via SDASM)*

74 XB-46 Jet Bomber - Model 109 (1944)

Front, three-quarter view from above shows the clean elegant lines of the XB-46. Like the B-45 before it, however, as handsome as the design may have been, it simply could not compete with swept-wing aircraft like Boeing's revolutionary six-engine XB-47 Stratojet that offered 600-mph speed. (National Archives via Dennis R. Jenkins)

In early 1944 the Army Air Force was eager to initiate bomber programs that would capitalize on the turbojet engine technology that was beginning to make significant progress. Requirements were developed for several classes of aircraft and a specification was issued in September for a medium jet bomber with a gross weight of 80,000 to 90,000 pounds, a range of 3,000 smi, and a speed of 500 mph. Convair submitted a predesign package on 6 November 1944 for its four-jet bomber design and received a letter contract in January 1945 for the initial Phase 1 design effort of the XB-46 (MX583). Three prototypes were ordered on 27 February and the contract was finalized on 7 June 1945. The design of the prototypes was nearly identical to the preliminary design information with the exception that the engine inlets were revised. The competitor for this medium bomber mission was the North American B-45. North American had received its contract a little earlier in 1945.

The Convair entry was powered by four GE TG-180

(J35) axial-flow jet engines; it had a gross weight of 91,000 pounds and a crew of three. Its wingspan was 113 feet and it was 105 feet 9 inches long. It had very clean lines, a graceful appearance with the long slim fuselage, and if visual appearance had been the criteria, it would have been the winner. As it turned out Convair's design was on the high side of the gross weight requirement range of 90,000 pounds and North American's B-45 was on the low side at 80,000 pounds. This resulted in the B-45 being a little faster with a little longer range using the same four engines. The XB-46 was thus outflanked by the B-45 on the down side as it was lighter and faster and by the Boeing B-47's superior performance on the up side.

At that time the Air Force was under heavy budgetary pressure and the prevailing opinion was to cancel the XB-46 and transfer the funds to the XA-44 program at Fort Worth. The tactical ground-support mission seemed to outweigh that of the medium bomber in many

Bomber Programs

The initial pre-design data package for what was to become the XB-46 was submitted on 6 November 1944. In very quick order a letter contract for the Phase 1 effort was issued on 17 January 1945, a mockup inspection was held on 29 January, and a contract for three prototypes issued on 27 February 1945. The XB-46 was a very clean and elegant design powered by four GE TG-180 turbojets. It had a high-aspect-ratio 1,285-square-foot wing with a 113-foot span. Slim fuselage was 105 feet 9 inches long and housed a crew of three. The aircraft had a gross weight of 90,000 pounds. *(Convair via SDASM)*

The XB-46 changed very little from the design concept initially submitted to the aircraft actually built. Most notable visual difference was the design of the engine inlet ducts. Unfortunately for the XB-46, it was overshadowed by both North American's B-45 Tornado that was smaller and lighter, and Boeing's swept-wing B-47 Stratojet with its greatly superior performance. Hence, the XB-46 was targeted for cancellation in October 1945. Convair resisted and convinced the Air Force to finish the first prototype and conduct the initial flight-testing. The XB-46 made its first flight on 2 April 1947 at San Diego but was scrapped soon after the flight test program was concluded. *(Convair via SDASM)*

North American Aviation's B-45 Tornado was America's first operational jet bomber, and saw action during the Korean War. As with many of the Convair proposals of this time period, the B-45 married a conventional straight-wing configuration airframe with turbojet powerplants. *(National Archives via Dennis R. Jenkins)*

quarters. Convair, in the midst of postwar downsizing, fought to keep the XB-46 program going and finally convinced the Air Force to continue and complete the first aircraft and to conduct early flight testing. In October 1945 the second and third XB-46s were canceled and the residual funds from the XB-46 were transferred to the XA-44. This latter program was to have been moved to San Diego but that did not happen and it, too, ended up being canceled. The first XB-46 was indeed completed, although without any of the military equipment, and its first flight took place on 2 April 1947 at San Diego. Convair conducted about 27 hours of flight testing, the Air Force conducted 20 hours at Muroc, and more than

40 hours at Wright-Patterson Air Force Base (WPAFB) assessing stability and control characteristics.

A brief study of a Photoreconnaissance version of the XB-46 was undertaken in May 1947. In this design most of the military equipment was removed and a camera suite, photoflash bombs, and a fourth crewmember were added to operate the photo equipment. Apparently this study generated no interest in view of the fact there was not going to be a production program for the bomber. The XB-46 met its ignominious end in 1952 when the remaining airframe portions were scrapped. The total cost of the program was about $4.8 million.

Bomber Programs

A design study was completed in March 1945 for a four-engine turboprop heavy bomber for the Air Force. It is believed to have been in response to a proposal request or a design study to requirements provided by the customer. Those requirements called for a 5,500-mile range with forty 500-pound bombs and a maximum speed of 475 mph. The Convair design was a large aircraft with a 2,000-square-foot wing, and a gross weight of 175,000 pounds. Four Wright GTAA regenerative 5,000-hp turboprop engines powered it. (Convair via SDASM)

This turboprop heavy bomber is of quite conventional design with straight wings accommodating the four turboprop engines. These 5,000-hp engines drove six-blade, counter-rotating, 15-foot propellers. With a wingspan of 141 feet 8 inches and a length of 145 feet 4 inches, this 175,000-pound gross weight airplane had an empty weight of 90,430 pounds and could carry up to a maximum of a 44,000-pound special weapon. (Convair via SDASM)

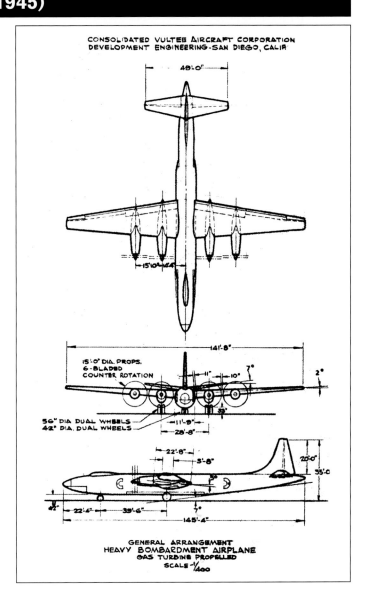

Toward the end of World War II, the Army Air Force initiated a series of requirements and design studies to exploit the rapidly developing gas turbine propulsion technology. This included studies of second-generation systems to meet the intercontinental requirement to be fulfilled by the B-36, as well as the medium-bombardment experimental programs that yielded the XB-45, XB-46, XB-47, and XB-48 aircraft. The second-generation heavy-bomber studies conducted by both Boeing and Convair tended to center on turboprop solutions.

Convair's turboprop bomber study, completed in March 1945, proposed a design that had the lines and slim look of the XB-46. It was a large, straight-wing configuration powered by four Wright GTAA regenerative engines rated at 5,000 hp, and it had a gross weight of 175,000 pounds. It was to carry 20,000 pounds of bombs for a maximum range of 5,670 smi at a cruise speed of 416 mph. It had a maximum speed of 477 mph and carried a flight crew of nine.

With both Convair and Boeing, the studies fell short of providing significant improvements over the B-36, especially after it had been enhanced with the addition of the four J47 turbojets in wingtip pods. The Air Force then turned to the turbojet solution for this mission, resulting in the B-52 heavy bomber, and the turboprops were out of contention for the heavy-bomber role.

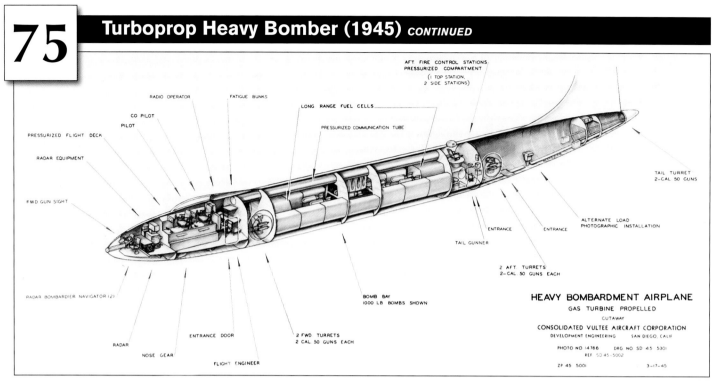

Defensive armament is shown on the cutaway view which includes a tail turret and fore- and aft-mounted fuselage low-profile turrets on each side. The flight crew was accommodated on the forward flight deck and a fire control station in the aft fuselage. Of note, this design provides for fuel cells along the fuselage sides of the long bomb bay. The side of the fuselage was faired outward to accommodate these fuel tanks and the fairing was terminated both fore and aft with the flush gun turrets. (Convair via SDASM)

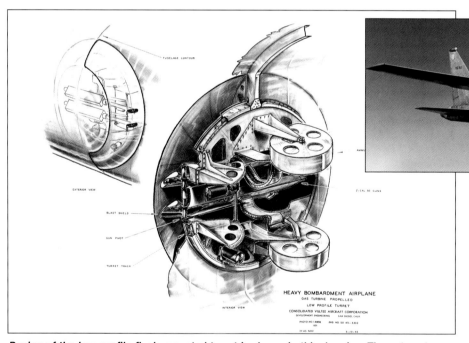

Design of the low-profile flush-mounted turret is shown in this drawing. These turrets were located at the fore and aft of the fuel cell section of the fuselage that lies adjacent to the bomb bay area. The four turrets as well as the tail turret each house two .50-cal. guns. Fuselage guns had 800 rounds of ammunition and the tail gun had 1,000 rounds. (Convair via SDASM)

The ultimate in jet bomber design of this era was Boeing's eight-engine B-52 Stratofortress which replaced Convair's prop-and-jet-powered B-36 Peacemaker as the backbone of U.S. strategic bombers in the mid-1950s. The XB-52 prototype is shown here, a design that evolved from a turboprop concept as well. Although the B-36 was the Air Force's longest-range aircraft up to that time, the new pure-jet B-52, with its inflight refueling capability, could literally fly anywhere in the world non-stop. (National Archives via Dennis R. Jenkins)

Bomber Programs

NAVY SEARCH LANDPLANE

CONSOLIDATED VULTEE AIRCRAFT CORPORATION

Convair's design to satisfy the Navy's April 1946 requirement for a High Performance Patrol Landplane was a turboprop plus jet-powered straight-wing airplane with a gross weight of 167,000 pounds. Six Westinghouse 25-D turboprops and four Westinghouse X24C turbojets powered it. The latter turbojets were used to provide thrust augmentation to assist with takeoff and for high speed. (Convair via SDASM)

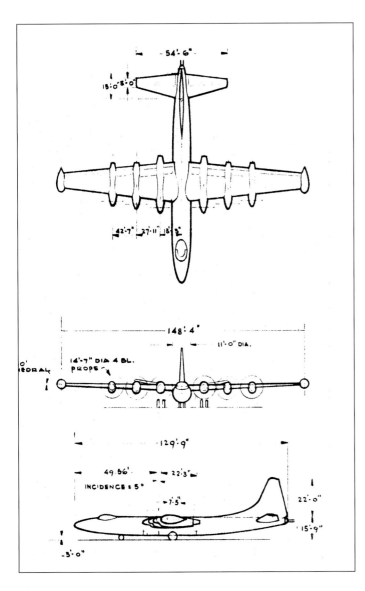

Convair conducted a study of a High Performance Patrol Landplane in April 1946 for the Navy, in response to a BuAer requirement document Aer-E-14HWK C-00722 dated 21 January 1946. It is not known if this requirement was a precursor to an actual procurement or was a study to further define the state of the art. The mission of this aircraft was to search ocean areas at high altitude and high speed, thus reducing its own vulnerability. Secondary missions were for photographic reconnaissance and for the delivery of special (nuclear) weapons.

The Convair design to satisfy this mission was a large straight-wing aircraft with 10 engines, six turboprops for climb and cruise, and four turbojets for takeoff and high speed. Top speed was 485 mph at 25,000 feet and combat radius was 1,650 miles.

An alternate configuration for this mission was also offered that used four Wright GTC turboprops rated at 10,000 hp. The physical characteristics and performance of the aircraft were essentially the same except the maximum velocity was 20 mph faster than the Westinghouse-powered version.

This aircraft had a wingspan of 148 feet 4 inches, a wing area of 2,200 square feet, and a length of 129 feet 9 inches. The turboprop engines were to produce 3,300 hp each, plus 990 pounds of thrust coupled with the X24C jet engine's 3,000 pounds of thrust. The engines were located at the rear of the nacelles and the turboprops were driven via long extension shafts. The jet engines were located in the intermediate and outboard nacelles and mounted on top of the wing with the turboprops beneath. (Convair via SDASM)

Bomber Programs

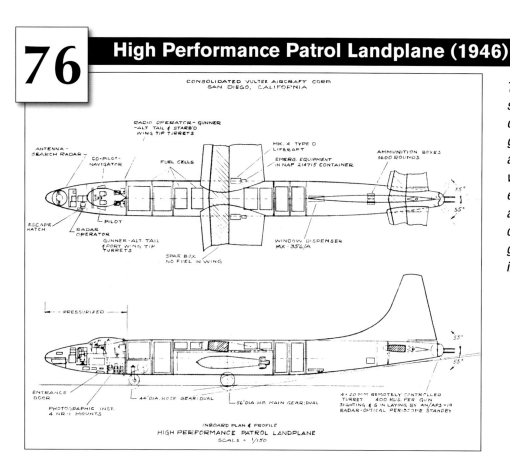

CONSOLIDATED VULTEE AIRCRAFT CORP.
SAN DIEGO, CALIFORNIA

INBOARD PLAN & PROFILE
HIGH PERFORMANCE PATROL LANDPLANE
SCALE = 1/150

The crew of five was located in the nose section and included, pilot, copilot, radar operator, radio operator/gunner, and tail gunner. Defensive armament consisted of a tail turret with four 20mm guns and two wingtip pod turrets with two 20mm guns each. The fuselage bomb bay could accommodate up to 24,000 pounds of ordnance. All of this aircraft's fuel (11,330 gallons) was also accommodated entirely in the fuselage. (Convair via SDASM)

77 Carrier-Based Bomber (1946)

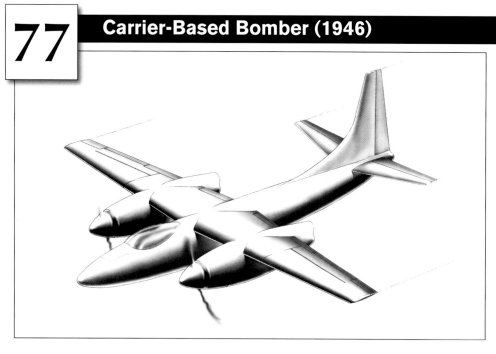

Proposal document that included these photos of Consolidated's Carrier Based Bomber was dated 19 April 1946. It was submitted in response to the Navy's OS-106 requirements and was a competitor in the competition won by North American with its AJ-1 Savage in June 1946. Two Pratt & Whitney R-2800 piston engines and two Westinghouse 24C turbojets powered Convair's design. (The AJ-1 had the same piston engines and a single fuselage-mounted J33 jet engine.) (Convair via SDASM)

After World War II was concluded, the Navy, in a classic interservice rivalry dispute, vigorously contested the Air Force monopoly on the atomic bomb and its delivery system. The Navy argued the B-29 was too limited in range and the projected B-36 was vulnerable without forward-based fighters. The Navy further contended that the mission could better be carried out with jet bombers based on the newly proposed super carriers.

Bomber Programs

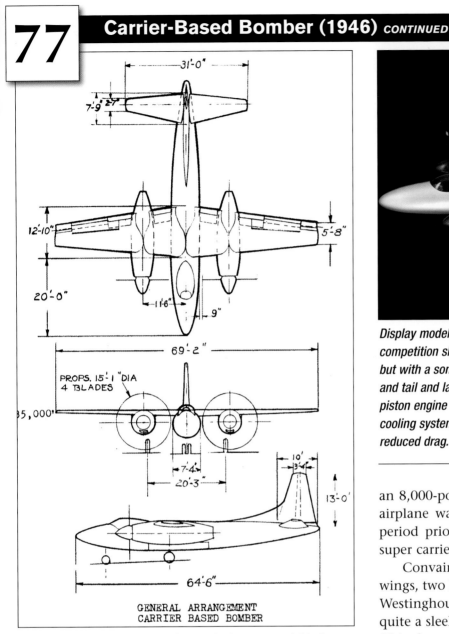

GENERAL ARRANGEMENT
CARRIER BASED BOMBER

Consolidated's design for this airplane had a gross weight of 44,372 pounds, wing area of 640 square feet, and span of 69 feet 2 inches. It carried a crew of two. This carrier-based airplane's mission was to deliver an 8,000-pound nuclear weapon from 35,000 feet and at least 300 miles range from the point of takeoff. Top speed of this design with military power was 485 mph at that altitude. (Convair via SDASM)

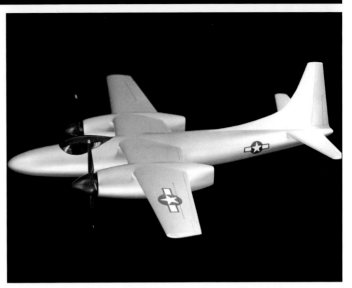

Display model of Consolidated's entry in the Carrier Bomber competition shows an airplane generally similar in size to the AJ-1, but with a somewhat cleaner configuration. It had a straight wing and tail and large, very streamlined nacelle that included both the piston engine and the turbojet. The augmented exhaust-thrust cooling system provided for a nacelle design that significantly reduced drag. (Convair via SDASM)

The Navy's first program to field a new nuclear-strike-capable carrier bomber was initiated in August 1945. A Request for Proposal was issued together with the requirements (Outline Specification OS-106), for the procurement of a carrier bomber capable of carrying an 8,000-pound weapon and a range of 300 nmi. This airplane was to provide an interim capability in the period prior to the availability of the projected new super carrier and an associated jet bomber.

Convair's proposal featured a design with straight wings, two Pratt & Whitney R-2800s, and two auxiliary Westinghouse J40 turbojet engines. Visually it was quite a sleek design, as was North American's proposed AJ-1, the eventual winner. This slim design was revised with a fuselage enlargement of the AJ-1 necessary to carry an early-design 10,000-pound atomic bomb. Convair's entry carried a crew of two, and had no defensive armament.

The proposal responses were submitted in May 1946; North American was declared the winner in June, and a contract was awarded for three XAJ-1 Savage prototype airplanes on 24 June 1946. The AJ-1 had the same piston engines as the Convair design but had a single fuselage-mounted J33 auxiliary jet rather than the two J40s. It is not known what Convair's shortcomings were but the lack of carrier aircraft experience was undoubtedly a contributing factor.

Bomber Programs

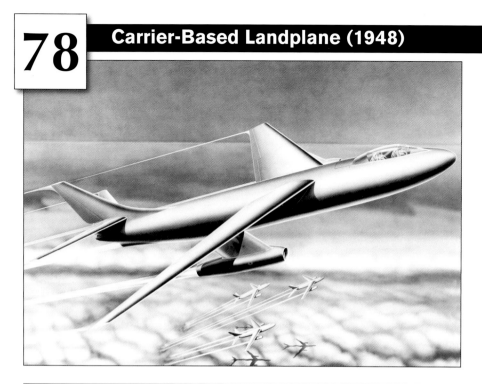

Convair's proposed entry in the Navy's competition of March 1948 for a Carrier Based Bomber for the new VCB Class super carrier featured three Westinghouse J40-WE-10 turbojets. The Convair design was larger and heavier than the winning Douglas A3D Skywarrior design and was right at the not-to-exceed maximum gross weight of 100,000 pounds and a design weight of 80,200 pounds. Douglas Aircraft Company, which had been studying the requirement for more than two years, was able to offer a smaller airplane with a gross weight of 70,000 pounds. The A3D went on to serve for nearly four decades, and was the largest airplane to ever operate routinely from an aircraft carrier. *(Convair via SDASM)*

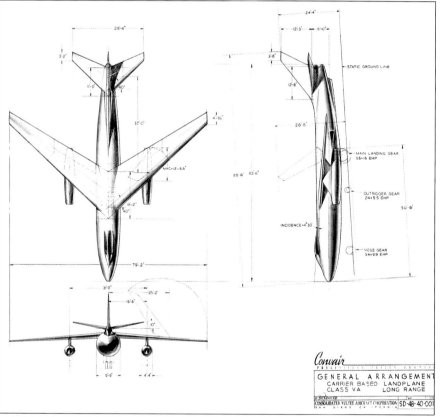

This airplane had two wing-mounted podded engines with the third engine installed in the aft fuselage. An alternate engine installation was offered where the pod-mounted engines were moved to a shoulder-wing mounting location. It had a bicycle-type main landing gear with outrigger that retracted into the engine pods. Wing area was 900 square feet (versus 779 square feet for the A3D) and the span was 79 feet 2 inches. The bomb load requirement called for a 10,000-pound weapon 16 feet by 60 inches in diameter. A gun turret with two 20mm guns was located in the tail. *(Convair via SDASM)*

The Navy had been considering jet bombers associated with super carriers since 1947, or possibly earlier, at a time when the Air Force lay claim to be the sole nuclear strike force for the military. The Navy lost the battle with the Air Force for the moment when its new super carrier the USS *United States* was canceled by Secretary of Defense Louis Johnson in April 1949, only days after the keel was laid. At that point the Navy

Bomber Programs

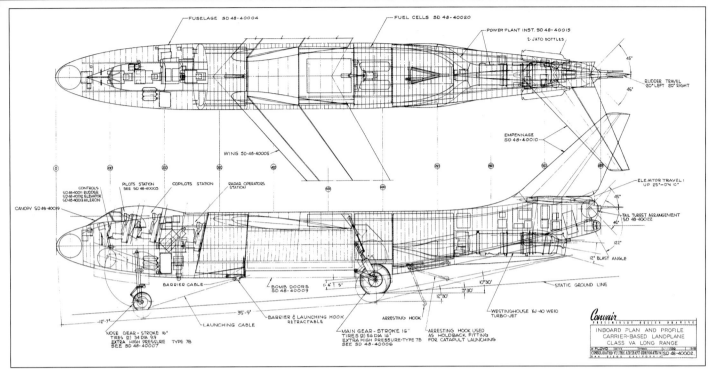

Landing gear retraction, aft engine installation, and tail gun installation are shown in this inboard profile. The three crewmembers were housed in the forward fuselage. Two JATO bottles were accommodated on the aft fuselage just outside the jet engine and were used to provide additional thrust for takeoff. The carrier-landing arrestor hook was located immediately behind the main landing gear. (Convair via SDASM)

was relegated to its current aircraft carrier modification-and-update program necessary to accommodate the interim piston-engine carrier bomber, the North American AJ-1 Savage.

The Navy's earliest studies had indicated that a jet bomber with a range of 1,500 nmi and carrying a 10,000-pound bomb would have a gross weight of up to 130,000 pounds. The USS *United States* was specifically designed to accommodate jet bombers of the 100,000-pound class. The Navy continued to push for this concept as a Navy nuclear role and initiated a competition for a 100,000-pound jet bomber in August 1948. Proposals were received for this procurement in December from six companies, including Convair.

Convair's design submitted in this proposal was a swept wing and tail configuration with three Westinghouse turbojets and a crew of three. Two of the jets were pod mounted on the wing and the third engine was installed in the aft fuselage. An alternate configuration was also submitted where the pod-mounted engines were moved to a fuselage shoulder location.

The airplane had a span of 79 feet 2 inches and overall length of 89 feet 6 inches. The gross weight was exactly 100,000 pounds. It had a bicycle landing gear with an outrigger on each side that retracted into the engine wing pods. A twin 20mm gun turret was included in the aft fuselage tail section. Maximum speed was 588 knots at sea level and the combat radius was 1,700 nmi.

Douglas and Curtiss were, however, the winners of this competition and were awarded a three-month study to refine their submittals. Douglas received the final nod with a Letter of Intent for Phase 1 & 2 and two XA3D-1 Skywarrior aircraft on 31 March 1949. Douglas had worked closely with the Navy on this requirement since 1947 and had already accomplished a large amount of design and analyses. The design was significantly smaller and well below the 100,000-pound limit, contributing greatly to its success. The Navy, meanwhile, was having better luck with its quest for a nuclear role when it was able to order a new super carrier, the USS *Forrestal*, in July 1951.

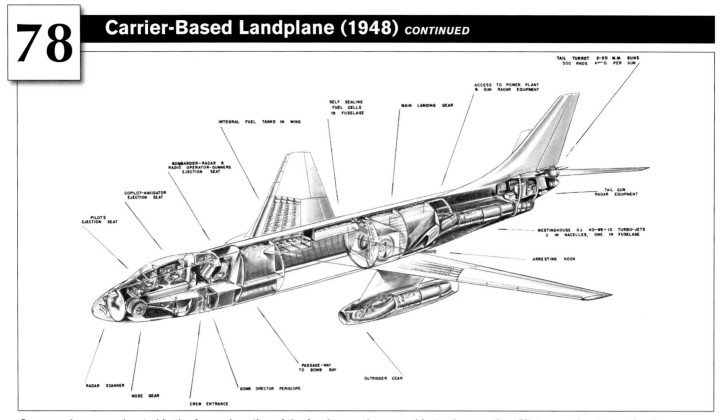

Crewmembers were located in the forward section of the fuselage and arranged in tandem seating. Pilots were located under the canopy and the bombardier radar operator was located in the fuselage immediately behind the canopy. All were equipped with ejection seats. The crew entrance was via the nose-gear wheel well adjacent to the bomb director periscope or optical bombsight. (Convair via SDASM)

The End Of The Bombers At Convair San Diego

In the end, most of the bomber engineering, study, and proposal effort was transferred to the Convair Fort Worth Division, and after 1948 only occasional support was provided by San Diego. The last large program having to do with bombers at the San Diego Division was the assist to the Fort Worth Division for modification of a number of B-36s. This work involved the addition of underwing-mounted jet engine pods and other modifications to 54 B-36s from April 1950 to February 1952.

Perhaps one of the most significant aircraft to emerge from Convair's advanced designs in addition to the World War II B-24 Liberator, was the B-36 Peacemaker, also referred to by its crewmembers as the "Magnesium Overcast." The B-36 served during the early years of the Cold War with the Strategic Air Command as the world's first true intercontinental bomber, and made the claim of successfully projecting the Air Force's global might without ever firing a shot in anger. (National Archives via Dennis R. Jenkins)

Bomber Programs

GLOSSARY

ARDC	Air Research and Development Command	GALCIT	Cal Tech	GETOL	Ground Effects Takeoff or Landing
ANP	Aircraft Nuclear Propulsion	CFR	circulating fuel reactor	L/b	length-to-beam ratio
ASW	anti-submarine warfare	CCR	compact core reactor	NACA	National Advisory Committee for Aeronautics
A/R	aspect ratio	CAC	Consolidated Aircraft Corporation	NYRBA	New York, Rio, Buenos Aires Airlines
AST	Assault Seaplane Transport	CVAC	Consolidated Vultee Aircraft Corporation	PAA	Pan American world Airways
ATC	Approved Type Certificate	DAC	Direct Air Cycle	SDASM	San Diego Air and Space Museum
AVCO	Aviation Corporation	ERCO	Engineering and Research Corporation	STOL	short takeoff or landing
BLC	Boundary Layer Control	GEBO	Generalized bomber	SPAWAR	Space & Naval Warfare Systems Command
Buweps	Bureau Of Weapons	GD-SD	General Dynamics Convair Division	VTOL	vertical takeoff or landing

SELECTED SOURCE MATERIAL

The majority of the factual information and graphics contained in this volume (probably upwards of 90 percent) is from the Convair archives located at the San Diego Air & Space Museum and are from the documents individually listed below. The documentation referenced represents proposal brochures, design study reports, and other reports aimed at customer relations and reporting technical investigations from the late 1930s to the mid 1960s. Many of the documents are brochures for the customers, promoting the subject airplanes. This archival material was donated by General Dynamics at the time of the dissolution of the San Diego Division in 1995.

Books

Jacobsen, Meyers K. *Convair B-36*, 1997, Schiffer.
Jenkins, Dennis R. *Magnesium Overcast*, 2001, Specialty Press.
Wagner, Ray. *American Combat Planes of the 20th Century*, 2004, John Bacon & Company.
Wagner, William. *Reuben Fleet*, 1976, Aero Publishers.
Wegg, John. *General Dynamics Aircraft*, 1990, Naval Institute Press.
Wolf, William. *Consolidated B-32 Dominator*, 2006, Schiffer.
Yenne, Bill. *Into the Sunset: The Convair Story*, 1995, The Greenwich Publishing Group, Inc.

Reports

All reports were generated by Consolidated Vultee Aircraft Corporation (CVAC) or Consolidated Aircraft Corporation (CAC).

XP3Y/PBY Catalina
Amphibian—2 Engine, ZP-28-001, 16 May 1940, CAC
Patrol Bomber Flying Boat—4 Engine, ZP-34-007, 24 August 1942, CAC

Model 28 Military Studies
Amphibian—2 Engine, No Report Number, 1 July 1939, CAC

Model 28 Commercial Studies
Commercial Flying Boat (PBY Type)—ZP-LB-023, No Date, CAC
PBY Commercial—ZP-BC-008, 1938, CAC

Model 29 Military Studies
Patrol Bomber Flying Boat—2 Engine, ZP-BM-023, 11 March 1940, CAC

Model 29 Commercial Studies
Flying Boat—4 Engine, ZP-BC-012, 17 November 1939, CAC

XPB3Y-1
Long Range Patrol Bomber (Plan II)—ZP-BM-006, No Date, CAC
Patrol Bomber Flying Boat—4 Engine, ZP-34-006, 16 February 1942, CAC

Model 34 Commercial Studies
Model 34 Commercial—ZP-34-003, 28 November 1941, CAC

Trans-Oceanic Flying Boat
Four Engine Commercial Trans-Oceanic Flying Boat (pages only), No Report Number, 7 August 1938, CAC

100-Passenger Seaplane
100 Passenger Flying Boat—ZP-BC-006, 1936, CAC
Four Engine Flying Boat—100 Passenger, ZP-BC-001, 1936, CAC
Four Engine Flying Boat—100 Passengers, ZP-SC-006, 1936, CAC
Four Engine Seaplane—ZP-SC-001, No date, CAC
Four Engine Seaplane—100 Passengers, ZP-SC-001, 12 January 1936, CAC

Prewar Two- and Three-Engine Flying Boats
Class VPB 2-Engine Flying Boat—1,650 sq ft Wing, No Report Number, 15 February 1938, CAC
Class VPB 3-Engine Flying Boat—1,500 sq ft Wing, No Report Number, 15 February 1938, CAC
Flying Boat Commercial—3 Engine, ZA-31-002, 29 September 1938, CAC
Flying Boat—3 Engine, ZA-31-002, No Date, CAC

Prewar Four-Engine Flying Boats
Flying Boat—4 Engine, 38 Night Passengers, ZA-BC-004, 9 September 1938, CAC

Model 31
Patrol Bomber Flying Boat—2 Engine, ZP-BM-018, 11 March 1939, CAC
Patrol Bomber Flying Boat—2 Engine, ZP-BM-019, 3 November 1939, CAC
Patrol Bomber Flying Boat—Medium Range, ZP-BM-013, February 1939, CAC
Patrol Bomber Flying Boat—Model 31, ZP-BM-022, 11 March 1940, CAC
Patrol Bomber Flying Boat—Model 31, ZP-BM-025, 16 April 1940, CAC
VP Flying Boat—2 Engine, ZP-31-007B, 27 July 1942, CAC

Model 31/XP4Y-1 Corregidor and Military Studies
V-P Type Flying Boat—P4Y-1, ZP-31-010, 14 December 1942, CAC

Model 31 Commercial Studies
Air Yacht—No Report Number, 21 December 1938, CAC
Flying Boat—Model 31 Commercial, ZP-BC-015, 1939, CAC
Flying Boat—Model 31 Commercial, ZP-BC-15, 21 April 1939, CAC
Flying Boat—Model 31 Design Study, No Report Number, 17 August 1938, CAC

Postwar Seaplane Designs
High Performance Flying Boat (Pre-Study)—No Report Number, Undated, CVAC
High Speed Boat (6 Jet)—No Report Number, No Date, CVAC
Patrol Reconnaissance Airplane—Single Float, ZP-44-15002, 2 December 1944, CVAC
Patrol Reconnaissance Airplane—Single Hull, ZP-44-15003, 2 December 1944, CVAC
Patrol Reconnaissance Airplane—Twin Float, ZP-44-15001, 18 November 1944, CVAC

P5Y
Model XP5Y-1 Patrol Flying Boat (Pre Design Sketchbook)—ZP-46-2001, 1 December 1946, CVAC
Patrol Seaplane—4 Engine (R-2800s), No Report Number, 20 January 1945, CVAC
Patrol Seaplane—4 WAC Turbines (Alternate), ZP-44-15005, 7 February 1945, CVAC

Flying Wing (Army and Navy)
Commercial Transport Type—Flying Wing, No Report Number, 1 January 1938, CAC
Landplane Bombardment Type—Flying Wing, No Report Number, 1 June 1938, CAC

Prewar One-, Two-, and Three-Engine Bombers
Attack Bomber—ZP-LB-013, 1938, CAC
Attack Bomber—2 Engine, ZP-LB-016, 17 January 1939, CAC
Bombardment Aircraft—2 Engine, ZP-LB-026, 19 January 1940, CAC
Bombardment Airplane—2 Engine, ZP-LB-027, 1 February 1940, CAC

B-24 Liberator
Bombardment Airplane—4 Engine, ZP-LB-023, 1939, CAC

C-87 Liberator Express and Model 32 Transport Studies
B-24/C-87 (Transport Versions)—ZP-32-012, 16 February 1942, CAC
B-24 Type Cargo Airplane (Conversion)—ZP-32-018, 9 November 1942, CAC
Four Engine Transport—ZP-32-007, 15 October 1941, CAC

Model 32 Commercial Studies
Commercial Landplane—4 Engine, ZP-32-006, 25 April 1941, CAC
Commercial Transport—4 Engine Model 32, ZP-32-009, 2 December 1942, CAC

B-32 Dominator
Bombardment Airplane (Heavy)—4 Engine, XB-32, ZP-33-003, 17 December 1940, CAC
Bombardment Airplane (Heavy)—4 Engine, ZP-33-002, 4 December 1940, CAC
Bombardment Airplane (Heavy)—4 Engine, ZP-33-004, 25 April 1941, CAC
Heavy Bombardment Airplane—B-32 (4 TG-100), ZP-44-12001, 20 October 1944, CAC
Model 33 Composite Designs (Bomber/Escort/Cargo)—ZP-33-015, 6 July 1942, CAC
Reconnaissance Landplane—4 Engine, ZP-45-10-001, 20 October 1944, CAC

Model 33 Transport Studies
Model 33 Cargo Airplane—ZP-33-025, 10 April 1943, CAC
Model 33—Four-Engine Cargo Transport, ZA-33-028, 1 September 1943, CAC
Model 33 Freighter—4 Engine Landplane, ZP-33-008, 11 September 1941, CAC

Model 33 Commercial Studies
Commercial Transport—4 Engine Landplane, ZP-33-010, 2 December 1941, CAC
Model 33 Commercial—4 Engine Landplane, ZA-33-007, 20 August 1941, CAC

B-36 Peacemaker
Heavy Bombardment Aircraft—6 Engine, ZP-36-005, 18 February 1942, CAC
Heavy Bombardment Aircraft—6 Engine, ZP-36-006, 21 February 1942, CAC
Heavy Bombardment Airplane—6 Engine, ZP-30-003, 17 December 1941, CAC
Heavy Bombardment Airplane—6 Engine, ZP-36-006, 3 April 1942, CAC
Heavy Bombardment Airplane—6 Engine, ZP-36-010, 18 January 1943, CAC

XC-99
100,000 Lb Cargo and Troop Transport (Proposal)—ZP-46-11,000, 20 August 1946, CVAC
Cargo Troop Transport—ZP-36-002, 30 January 1942, CAC
Model 37 Composite—ZP-37-002, 25 June 1943, CVAC
Six Engine Transport XC-99—ZP-37-003, 3 May 1944, CVAC
Transport Airplane—6 Engine, Model 37, ZP-44-5002, 29 May 1945, CVAC
XC-99 Cargo Transport—ZP-37-001, 8 October 1942, CAC
XC-99—No Report Number, 1943, CAC

Flying Wing B-36 Comparison
Heavy Bomber Flying Wing—6 Engine, ZP-XM-001, 8 May 1942, CAC

PB4Y Privateer
Four Engine Reconnaissance Landplane—ZP-100-002, 26 July 1943, CAC
Four Engine Transport Airplane—RY-3, ZP-101-003, 1 July 1944, CAC

Tailless Twin-Engine Patrol Studies
Tailless vs. Conventional—1 Engine Patrol Plane, ZP-015, 16 October 1943, CVAC
Twin Engine Bombardment Airplane—Tailless Type, ZP-020, 3 December 1943, CVAC
Two Engine Patrol Landplane (Revised)—ZP-029, 8 April 1944, CVAC
Two Engine Patrol Landplane—Tailless Type, ZP-017, 3 December 1943, CVAC
Two Engine Patrol Landplane—Tailless, ZP-017, 3 December 1943, CVAC

Conventional Twin-Engine Patrol Study
Tailless vs. Conventional—1 Engine Patrol Plane, ZP-015, 16 October 1943, CVAC
Two Engine Patrol Landplane (Conventional)—ZP-013, 8 October 1943, CVAC
Two Engine Patrol Landplane—Tailless Type, ZP-010, 20 August 1943, CVAC

Tailless Four-Engine Army Bomber
Four Engine Bombardment Airplane—Tailless Type, ZP-009, 18 August 1943, CVAC
Tailless vs. Conventional—4 Engine Bombardment Airplane, ZP-014, 16 October 1943, CVAC

Twin-Engine Attack Airplane
Twin Engine Attack Airplane—ZP-008, 20 July 1943, CAC
Twin Engine Attack and Light Bombardment Airplane—ZP-001, 7 April 1943, CAC

XB-46
Medium Bombardment Aircraft—Jet Propelled, ZP-44-17001, 23 October 1944, CVAC
XB-46 Medium Bombardment Airplane—ZP-44-17002, 5 January 1945, CVAC

Turboprop Heavy Bomber
Heavy Bombardment Airplane (Gas Turbine)—ZP-45-5001, 21 March 1945, CVAC

High Performance Patrol Landplane
Navy Search Landplane—ZP-46-8000, 19 April 1946, CVAC

Carrier-Based Bomber
Carrier Based Bomber (Class VB)—ZP-46-6000, 19 April 1946, CVAC
Carrier Based Bomber—Model Photographs, No Report Number, 1 April 1946, CVAC